GUIDELINES FOR GREAT FRAME DESIGN

GREG PERKINS

HOW TO ENHANCE YOUR HOME OR OFFICE WITH FRAMED ART

PFM BOOKS
MORGANVILLE, NEW JERSEY

First Paperback Printing 2009

ISBN: 978-0-9663189-2-0

Library of Congress Control Number: 2007938530

Printed in China

CONTENTS

IX FOREWORD

XI PREFACE

1 CHAPTER ONE: ABOUT CUSTOM FRAMING

7 CHAPTER TWO: THE DIFFERENT FACES OF CUSTOM FRAMING

13 CHAPTER THREE: CUSTOM FRAMING GUIDELINES

15 CHAPTER FOUR: FRAME MOULDING

43 CHAPTER FIVE: MATTING

81 CHAPTER SIX: GLASS & ACRYLIC

85 CHAPTER SEVEN: MOUNTING

89 CHAPTER EIGHT: PUTTING IT ALL TOGETHER

119 CHAPTER NINE: LOOKING FOR SOMETHING UNIQUE?

123 CHAPTER TEN: DISPLAYING FRAMED ART

125 A FINAL NOTE

129 INDEX

Photo by Clark R. Hill

FOREWORD

Throughout history, great art has been made even more impressive by great frames. As the French artist Edouard Manet once said, "That without the frame my art is nothing."

I have known and admired Greg Perkins for 20 years, and I consider him one of the foremost frame design experts in this country. He, more than anyone else I know, would agree with Manet's statement.

His book, *Guidelines for Great Frame Design*, has been a long time in coming, and there is no other book quite like it. It is the first publication that not only speaks to custom framers but also to designers, architects, and everyone else who wants to hang beautifully framed artwork in their homes and offices. This book invites us into the creative thought processes of a great frame designer like Greg, who makes the art of frame design seem easy.

Unlike other time periods, we live in a wonderfully eclectic age. In the past, frame design was essentially limited to the styles of the period in which the artwork was created. Today, the choices of what we display in residential or corporate environments is limited only by our imaginations, as you will see from the many wonderful examples presented in this book.

As Greg points out, working with a custom picture framer is the best way to achieve a look that exceeds your expectations. Like most professions, picture framing has a "language" of its own. When you learn that language, you gain a clearer understanding of the frame design process and all its possibilities.

In *Guidelines for Great Frame Design*, you will discover that there is no one right way to frame something. Greg's designs are masterful-sometimes classic and sometimes striking in their originality. Greg also provides numerous examples of the same piece framed in different ways, all with equally impressive results.

That's why Greg insists that he is providing us with guidelines and not rules. What guidelines do, he says, is to provide direction without inhibiting your creativity or personal style. For with the knowledge of good guidelines come great designs.

Bruce Gherman
Executive Publisher
Picture Framing Magazine

PREFACE

I love custom framing! If you aren't passionate about it now, I hope I'm able to change that as you make your way through this book. A lot of the framing hanging on walls of homes and businesses is so mediocre that you may not have had many opportunities to get excited by it. Some framed art simply fills space. It's easy to not even notice some of it, let alone care about it. Good quality, well-designed custom framing enriches the rooms where it is displayed, and it can enrich the lives of everyone who lives with it.

If you already have great feelings about custom framing, I hope to further strengthen them and inspire you to want more of it. If you've never really thought about custom framing, don't know anything about it, or simply don't understand its value, I'd like is to open your eyes to its beauty and how it enhances the display of art.

The intent of this book is to make you comfortable with the process of custom frame design and to share framing ideas that you may not be aware of. Custom framing provides one of the best ways to personalize your home or business, and it gives you an opportunity to surround yourself with art and other things that can inspire you and make your life more enjoyable.

My interest in art and design started at an early age. As a child, I was fascinated with residential architecture. While other kids were reading comic books, I was buying house plan magazines. After studying them, I drew my own floor plans and the exterior elevations. I continue to be interested in home design today. In school I was interested in art classes more than anything else. As a teenager I also started working on art at home and later also began to do woodworking.

I became a retail picture framer in the early 70s, working at a do-it-yourself shop in Bloomington, Minnesota, called The Frame Factory. I was still in high school when I started there but had decided to attend the Minneapolis College of Art & Design the following year. Working in a frame shop seemed like it would be a good job until I found a "real" career. I had no idea my love for picture framing would grow so much that custom framing would become my lifelong career.

Eventually I opened Great Frames Gallery in Bloomington. It was then that my passion for creative frame design really took off. I entered numerous framing industry competitions, winning 25 awards for design and craftsmanship.

In 1987, while teaching framing workshops for the Professional Picture Framers Association at a convention in Toronto, I met Nancy Black, a frame shop owner from Arlington, Texas. Two years later we were married. A lot of people get bonuses from their jobs, but not everyone gets a spouse from it!

Around that same time I became an independent consultant to the framing industry and traveled extensively, educating other framers. It wasn't long before most of my time was spent working for Larson-Juhl, the world's largest manufacturer and distributor of custom picture framing products. In 1991 I was hired for a full-time position at the company headquarters in Atlanta, where I continue to work today.

My childhood self-training as a designer has paid off. I now design trade show exhibition booths and occasionally have opportunities to design retail frame shops for other people. In addition, I write articles for magazines and present seminars for retail framers, both domestically and internationally.

When time permits, I still enjoy creating art. My interests are all over the place, so my art is quite varied. I sometimes think of myself as "Psycho Artist" because my style is so fragmented. As you flip through the pages of this book, you will find a number of my pieces and see what I'm talking about.

After more than three decades in this business, it seemed like the right time to compile a lot of the information that I teach or write about for a book on the subject. It has been a rewarding project to design the framing and to write about it all. I hope you will enjoy it.

THANKS

I am listed as the author of this book, but projects like this aren't a solitary task. The words and ideas are the result of experiences from both my business and personal lives. This includes interactions with many other people. There would be no book without them.

SPECIAL THANKS TO...

Nancy Perkins: Thanks to my wife for encouraging me, not only during this project but also in everything I do. She is very tolerant of my workaholic nature and understanding of my business schedule. I dream that one day she and I will be able to relax, sitting side by side, listening to crashing waves and watching the sunset.

Ruby Perkins: Thanks to my mom for handing me a help-wanted ad for a picture-framing job over 30 years ago. I never could have imagined that moment would lead to this. Mothers are responsible for giving life to their kids, but my mom also set me up to live this specific life. For that, I am extremely thankful.

Larson-Juhl Friends: Thanks to my friends and co-workers at Larson-Juhl. I am inspired by their creativity and energy. I am proud to work with a group of people who excel at what they do and care deeply about making a positive contribution to the industry.

Framing Industry Friends: In the course of more than three decades in the framing industry, I have met many wonderful people and formed special friendships that transcend work. Thanks to all of them for contributing so much to my life.

PRODUCT CONTRIBUTIONS

Thanks also to the companies that contributed many of the beautiful products in this book. The biggest complement is to enjoy the art and framing you see in this book and want to purchase similar products. Most of them are available through custom frame shops across America.

Framing provided by:

Larson-Juhl (www.larsonjuhl.com)

Art provided by:

Artaissance (www.artthatfits.com)
Bruce McGaw Graphics (www.bmcgaw.com)
Deljou Art Group (www.deljouartgroup.com)
Designs Ink (www.designsinkart.com)
Editions Limited/Drybrush Graphics (www.editionslimited.com)
Grand Image (www.grandimage.com)
Linda Omelianchuk (www.omelianchuk.com)
New Era Publishing (www.newerahd.com)
World Art Group (www.oldworldprintsltd.com)
Progressive Fine Art (www.progressivefineart.com)
Rosenbaum Fine Art (www.rosenbaumfineart.com)
VINSONart.com (www.vinsonart.com)
Wild Apple Graphics (www.wildapple.com)

GUIDELINES FOR
GREAT FRAME DESIGN

Photo by Clark R. Hill

ARIAS FOR
SOPRANO
SCHIRMER OPERA ANTHOLOGY

CHAPTER ONE

ABOUT CUSTOM FRAMING

Custom framing is classified as a luxury product. To me, it's one of life's basic necessities. I'm exaggerating, but I do hate to imagine what it would be like to live without all the custom framed art and objects I have collected over the years. They bring me joy and add a lot to the quality of my daily experience. I can look at all of the custom-framed pieces around me, and most of them help me recall special people, wonderful places, or bring back some other significant memory. Even the things that have no deep personal meaning bring me happiness just because they're there.

Have you ever entered a fully furnished room and had the feeling that it still looks unfinished, cold, or impersonal? At least a part of the problem with those rooms is usually an absence of beautifully framed art. From now on, when you enter a lackluster room, take a careful look around; you'll probably see exactly what I am talking about. Walls without framed art are like large expanses of wood flooring without area rugs or a home without any landscaping. They look incomplete. A lack of framed art on the walls or even cheap, uninspired framing that just fills space can bring an entire room down a few notches. Quality custom framing provides one of the easiest and best ways to transform an ordinary-looking room into a more stylish space, one filled with personality.

Nicely framed art can help divert attention away from ordinary furniture or architecture. The added perception of quality that good framing brings to a room can lead people to believe that the quality of the entire decor is better than it may actually be.

Custom framing also has the power to help an already attractive room make an even stronger impression on the people who enter it. Framed art can breathe new life into ordinary decor by injecting interesting new focal points into the space.

For people who aren't knowledgeable about frame design or the process of custom framing, it's not unusual to either avoid it or to be uncomfortable dealing with it. For those who do understand it, however, it can be an enjoyable experience. As a group, professional custom picture framers love what they do and have a great time helping people with all their custom framing needs. People

Photo by Clark R. Hill

who purchase custom framing express a consistently high level of satisfaction with the framing they purchase, according to consumer research conducted by the Professional Picture Framers Association.

This book is designed to expand your knowledge of custom framing so that you'll feel more at ease in a frame shop. Once you develop a better understanding of frame design you'll find it faster and easier to make decisions, and the choices you make will be smarter. Chances are, you will be more satisfied with the results.

When you fill your home with beautiful custom framing, you can start every day looking at art or objects that mean a lot to you—not just something you bought off a shelf to match your color scheme. If life is good now, think of how much better it will be when you surround yourself with your own special, framed treasures.

THERE ARE NO RULES

There are no rules governing custom frame design. And it would be hard to impose any rules because there are so many variables for each item you frame. As a professional designer, I am thankful no one is telling me how to design. Back when I was a novice framer, however, I craved guidance that wasn't available. At that point, even generalizations would have been better than nothing. Even today there isn't much written on this subject.

In the early 1980s I started presenting framing seminars to my fellow custom framers. It became clear that many of the people who attended those sessions were hungry for better design information. Starting then, I decided that each time I created a new frame design, I would consciously keep track of why I made each decision. It didn't take long to see that my best designs were actually logical. I wasn't just picking mat colors and frame styles because I liked them. I quickly learned to study all the details of every piece of art and seek out the unique qualities of each one. Most of all, I let the art "tell" me what it needed. That may sound a little strange, but it works.

The best solution I found for bringing stronger foundations to my own frame design choices came when I wrote down the reasons I made each decision. I soon had lists of logical reasons for my design choices. I decided to call them Frame Design Guidelines. I have been using these tips for more than 25 years, and I have taught them to thousands of other framers, both domestically and internationally. And they really do work.

THERE ISN'T ONE CORRECT DESIGN

If 10 experienced custom framers were asked to frame the same print, you would see 10 different frame designs. There might be some similarities, but no two would be identical. If 10 other people were asked to choose their favorite from those

The gold color of this frame would show up well on a dark wall. The opulent pattern helps to make this a real focal point.

This choice would show up best on a lighter wall color. Less ornamentation on the frame helps make this design a bit more casual and contemporary.
("Paris Wine" by Vladan. Courtesy of Progressive Fine Art)

10 framed pieces, one may be more popular than the others but several would probably receive votes. The reason is simple: with frame design, there isn't a right solution. It's more a matter of personal taste. In fact, one of the main reasons that there aren't any firm rules for custom framing is that people have such varied tastes.

That doesn't mean there can't be wrong frame designs. These usually result from trying to match framing to a room rather than enhancing the art. Or, in an effort to save money, using too narrow a mat and moulding, which will visually cheapen the art a lot more than the dollars that were saved. When framing, the most important thing is to maintain the integrity of the art. Fortunately, that can usually be done in a number of different ways.

THE VALUE OF PURCHASING CUSTOM FRAMING

Custom framing provides the opportunity to make the best selections for every piece of art and every set of circumstances. The goal of custom framing is not to match room decor but rather to enhance the art in the location where it will hang. Being true to the art is the best way to create a design that will endure when you change your decor or move. Custom frame design allows you to choose all the framing materials for a piece of art so you end up with exactly what you want.

Quality custom framing helps present and protect your special treasures. This mask could hang directly on the wall, but the framing adds detail and importance that call attention to it.

You don't have to settle for something that is average, or even almost perfect, because a custom frame design can incorporate all of the elements that will make it totally perfect for you. When you choose classic versus trendy design, your framing will endure.

In addition to all the beautiful art you can frame, custom framing can help you preserve items of your family history, collectibles, or just about anything else you want to display. Here are just a few of the things you can frame:

- Family Photos
- Wedding Memorabilia (invitation, license, dried flowers, and the gown itself)
- Graduation Memorabilia (diploma, announcement, tassel)
- Military Memorabilia (medals, documents, uniforms)
- Sports Memorabilia (baseball cards, golf hole in one, jerseys)
- Vacation Memorabilia (photos, postcards, shells, rocks, ticket stubs)
- Hobbies (stamps, coins, matchbooks, buttons)
- Collectibles (plates, antique purses, jewelry, fountain pens)
- Needle Art (needlepoint, cross-stitch, embroidery)
- Textiles (christening gowns, kimonos, vintage clothes)
- Musical Instruments (violin, clarinet, flute)
- Documents (land deeds, calligraphy, awards)
- Children's Art (finger paints, drawings, art for kids)

Some people consider custom framing to be expensive. In fact, the dollars you invest in custom framing can provide years of joy that will far outlast most other ways in which you might spend the same amount of money. Many people wonder why custom framing can cost more than the print being framed. The fact is that much of the artworks being framed today are posters or decorative prints, which are simply pieces of paper run through a printing press as a mass-produced product. On the other hand, better quality artwork may, indeed cost more than the framing.

By comparison, custom framing is physically assembled one at a time by a person. It is built to your specifications and may use mouldings that were painstakingly hand-finished. You can think of custom framing as handmade furniture for the wall.

Custom products almost always cost more than off-the-shelf, readymade products. Investing in a custom product can also provide a much better long-term value. For example, a $39 pre-framed poster may seem like a great bargain when you purchase it. However, in a short time you may begin to see why it was so cheap. The inexpensive materials used for both the art and framing typically don't hold up well over time. The art itself may fade noticeably in less than a year due to low-quality inks. Lower-priced frames are also more prone to warping, problems with the finish changing, or corners coming loose from stress. Often, the prints are not properly mounted, leaving them free to buckle under the glass. All these problems lead to an unsightly appearance that sends the framed piece to a garage sale after a couple years.

Most reputable custom framers use higher quality products that hold up well for years to come. The frames themselves are more comparable in quality to good furniture. Frames may actually last longer than most furniture because a frame on a wall doesn't generally get as much abuse. An experienced craftsman frames each piece, bringing a higher level of quality and pride to the job than you see in most manufactured framing.

A quote from custom framer Ray Dwyer says it all: "Quality custom framing will outlast the carpeting and furniture in your home, as well as all the cars, clothing, sound systems, televisions, and kitchen appliances you'll ever purchase. It can pass down from generation to generation and still look as good as the day it was done."

A $200 investment in custom framing can provide a quality piece you can pass down to the next generation. If you keep the framing for just 10 years, your cost is only $20 a year. And custom framing can easily last 50 years if you take care of it. That brings the annual price down to just $4, which is just over a penny a day. Many people are still enjoying Victorian frames over 100 years old that their great-grandparents had hanging in their homes.

If you need further reassurance that custom framing is a remarkable value, consider the cost of other things you may spend money on. A $50 floral arrangement may look good for five days, meaning it costs $10 a day. A $100 meal at a restaurant probably lasts an hour and a half, so every minute of that experience costs you $1.11.

Another point to consider is that custom framing can protect and preserve things you already own. When your belongings are properly cared for, they won't lose their monetary or sentimental value. When family treasures are left unframed or when they are improperly handled, they can become irreparably damaged and will be lost forever.

Money spent on custom framing is money well spent. The phrase "You get what you pay for" definitely does apply to custom framing. Inexpensive framing, on the other hand, is rarely a bargain. If and when inferior materials are used, they seldom hold up over time, meaning they will have to be replaced sooner. When you are looking for a bargain, consider spending more initially to purchase something that will last far longer.

It's fun to frame something unexpected like these sand dollars and starfish. They can create a real conversation piece when hanging on the wall.

This just goes to show you that almost anything can look good in a nice frame. Who would think 23 pencils could become a work of art?

modernism
PIERRE CHAREAU
THE NEW AMERICAN APARTMENT

CHAPTER TWO

THE DIFFERENT FACES OF CUSTOM FRAMING

As with any other product you can purchase, everyone likes choices in custom framing. People have their own tastes and interests, leading them to different solutions. Location and purpose are also important. If a person buys custom framing or framed art to use in two different places or to give to two different people, he or she may approach each purchase with different criteria.

When you go to a museum and look at the paintings on display, the framing is usually more elaborate than what you'll see on prints of the same art in most frame shops, galleries, homes, or offices. To understand the reasons for this, it's important to look at some of the different framing philosophies that underlie the many types of framing looks.

FRAMES IN MUSEUMS

In a museum you are likely to encounter grander, more opulent frames than you'll see almost anyplace else. For one thing, the value of the art on display merits the higher level of framing. But there is a different reason that most, if not all, of the frames in a museum are chosen. Throughout history artists have designed their own frames or worked with individual frame makers to create signature looks for their works. In addition, specific frame styles have become known as "the" choice for art from particular countries, eras, or styles.

Museum curators are conscientious about using frames from the right period, bringing a higher level of authenticity to the framed art. Whether the frames are original or reproductions of the same style, they provide the desired look.

Many pieces of art in museums are in what are called finished corner frames. This term means that there are no visible seams at the corners because the finish was applied after the frame was assembled. Finished corner frames, also known as closed corner frames, are available in select frame shops and galleries. While there are cheap imitations, handcrafted finished corner frames are considered the best frames made.

Fine art deserves fine custom framing. This original art was framed with a 22 karat gold-finished corner frame, enhancing its beauty. Finished-corner frames like this are handcrafted to meet the needs of each customer. *("De la Mundo III" by Greg Perkins. Prints of this image available through Artaissance, www.artthatfits.com)*

Sometimes pairs or sets of art have completely different colors schemes from one image to the next. A good way to unify the grouping is with off-white mats. If that sounds a bit boring for your taste, you can have a bit of fun by customizing each piece with a different frame color that relates to the predominant color in the art. When using a variety of frame colors, look for frames of similar value so one frame doesn't overpower another.

FRAMES IN ART GALLERIES

Custom framing on display in art galleries is generally simple in its design compared to framing in a museum or frame shop. When I have asked gallery owners or artists why they choose the frames they do, the common answers are:

- The artists frame their own works inexpensively because they have a low budget.
- The frames are simple and unobtrusive to focus full attention on the art and promote a more unified look throughout the gallery.
- The frames are really just intended for the gallery presentation. The people who purchase individual pieces can re-frame them however they want to suit their own tastes.

This doesn't mean that the framing you see in a gallery is bad or incorrect. Some of it is exquisite. There are many times, though, when a standard frame selection used on an entire series of art in a gallery doesn't reflect the personality of each piece of art. If you purchase art from a gallery, you have the option to frame it differently for your home.

A white mat and white gold frame provide a sophisticated look for this piece of art. The clean lines of this presentation is similar to what is used in many galleries. ("Collision 2" Framing by Jay Hall. Courtesy of Rosenbaum Fine Art)

FRAMING FOR CORPORATE OFFICES/STORES/RESTAURANTS

"Commercial" framing varies more dramatically than framing in other categories. There are a number of reasons this is so. The common ones are:

- Budget
- Knowledge of Framing
- Purpose
- Interest in Framing

Here's how each of these reasons has an impact on the finished product.

Budget: A reasonable budget will buy good quality framing that looks great and with proper care will stay that way for many years to come. On the other side, a low budget may buy framing that may not look great when it is new and goes downhill from there. The difference between the beautiful framing in the lobby of an upscale hotel and the framing in a fast food restaurant is likely to be budget-related. For a design project, make sure the designer provides an appropriate budget for framed art. Some designers tend to place art too low in their priorities and end up without enough funds to purchase good quality art or framing. Inexpensive pieces may look okay for a short time, but most don't hold up well over the long haul.

Purpose: What is the purpose of the framed art? Pieces hanging in the prime areas of an office building often are higher quality than the pieces in bathrooms or a back hall. Also, there may be projects that are not intended to last for a long period of time. For example, framed architectural drawings and floor plans in the sales office of a new subdivision are discarded when the neighborhood is built out.

No knowledge: If those responsible for purchasing commercial framing have no knowledge of what they are buying, they may end up purchasing inferior products that do not hold up. For example, some frames are made of plastic but look just like wood when they are new. Age and conditions can cause plastic to warp, making unsightly problems on walls.

No interest: If a person purchasing framing and art for an office, restaurant, or other business has an interest in it, it will get more attention, including the details. The result will be framing that looks much better and is likely to last longer. Have you ever seen a wall in a doctor's office filled with diplomas that are buckled under the glass and surrounded by low quality frames? This is probably the result of no one having an interest. As an experienced framer I cringe when I see those hard-earned pieces of paper deteriorating due to improper handling. Without spending a fortune, diplomas can be framed with the appropriate materials and techniques to hold up for years to come.

Three mats and a double frame were used in this frame design. A pair of chopsticks were also glued to the frame to create a unique touch, coordinating with the Asian theme. ("Wisdom" by Chris A. Paschke. Courtesy of Designs Ink, www.designsinkart.com)

CUSTOM FRAME SHOPS

The framed pieces in a frame shop may be a mix of the other categories listed here to appeal to the broad audience that the business is trying to attract. Shops catering to an upscale audience usually frame more elaborately than a shop selling inexpensive posters. Framers may frame fine art differently than you would see it framed in a gallery. The simple frame designs in many galleries don't always reflect the individual personalities of each piece of art. Custom framing can bring out the best features in every piece of art. A good custom framer is also conscious of choosing frame designs that will enhance the art in the environment where it will be displayed.

FRAMING FOR THE HOME

The framing you see in your friends' and relatives' homes is probably as varied as any you'll see anywhere. The quality and design of framing usually depends on the owners' tastes and their interest in their homes. Other factors that alter people's framing choices include the value of their art, their knowledge of framing, and their financial situations.

Fine art should be framed with classic, dignified designs befitting the quality of the piece. Spending a bit more for a quality job can directly add to the potential future value of the art and the longevity of both the art and framing. Inexpensive prints are sometimes framed the same way as fine art, giving a casual viewer the impression that it is a better quality piece. When art is purchased because it matches the decor of a room, not because the buyer has a passion for it, the framing typically ends up being more decorative, also matching the room. Many people enjoy decorating their homes this way. While it may not maintain the integrity of the art, it is important to surround yourself with art you love and things that inspire you. Whether your budget allows for original art or less expensive posters, when you love the image and it is well framed, it will be something you can enjoy for a long time.

These three abstract paintings on paper are all by the same artist and are approximately the same size. Three completely different framing treatments show that there are numerous ways to design framing for similar styles of art. Some of the differences are the mat colors and textures, mat border proportions, and the moulding colors and styles. The point is to be open-minded about all the opportunities available. *("Midge Series" by Craig Alan. Copyright Deljou Art Group)*

CHAPTER THREE

CUSTOM FRAMING GUIDELINES

It can be challenging to make frame and mat selections if you don't have any direction. Frame Design Guidelines can provide that direction without inhibiting your creativity or personal style. When choosing how to frame something, it is easy to be impulsive and simply choose your favorite colors and styles. However, frame designs can quickly lose their effectiveness if those choices aren't well suited to the art.

So what exactly is a framing guideline? Unlike rules, guidelines are just recommendations. For example, one guideline is "When choosing mat colors, blend the mat to the background color in the art so the foreground details stand out."

Framing guidelines are like road maps. When you look at a map, you can generally go from point A to point B via several different routes. One is the most direct, another the most scenic, and yet another may be the longest but perhaps the quickest way. Having all those choices provides a number of valid options and the chance to pick the route that is best for you. Framing guidelines work the same way. Guidelines help you keep important considerations at the top of your mind so you can make smart decisions based on your particular needs.

Two guidelines may lead to different results, yet both choices may be equally effective. Ultimately, you still have to decide which of the guidelines will work best for the specific circumstances dictated by what you're framing and where you want to display a piece of art.

Different guidelines may also contradict each other from time to time. That isn't a problem because guidelines are just independent ideas. You can achieve good results with each guideline, but they may be very different from one another.

Guidelines also help by breaking the entire design process into smaller pieces. When you can take it step by step, using a logical process, each decision is easier to make. Once you have a chance to read the guidelines and see examples of them in action, the process of successful frame design will be quite clear.

A light neutral mat (top) blends with the background paper the art is printed on. This mat allows all of the colors to project equally. The gray mat blends with colors in the art but it contrasts so strongly with the white paper that the background becomes too much of a focus. ("Silhouette Chartreuse" by Duplock. Copyright King & McGaw. Screen printed in the U.K. at Artizan Editions.)

CHAPTER FOUR

FRAME MOULDING

The four sides of a frame are cut from lengths of moulding. There are thousands of different mouldings available in frame shops across the United States and in other countries. Choosing just one moulding can be time consuming and confusing if you don't know how to go about it. The key is to find a focus. It's like choosing one entree from an elaborate menu at a new restaurant. You can begin by choosing a category, such as seafood or chicken, then look at the remaining choices to select one that suits your tastes.

The three main points to consider when selecting frame moulding are color, style, and scale. I recommend narrowing down your choices in that order. Try to visualize a look that will work best for the piece that you want to frame. When you have a focus and quickly rule out options you know won't work, you eliminate the majority of choices, making it faster to zero in on what you like.

The Dutch master Jan Vermeer was known for the use of light in his paintings. A dark frame contrasts with the light radiating in the art, calling attention to the light source in the completed design. The dark color and aged patina of the frame provide a nice look that fits with the era of the art.
("The Artist's Studio" by Jan Vermeer)

MOULDING COLOR

A good place to start thinking about a frame is to determine the color that will suit your needs the best. For example, if you choose silver, you can eliminate all other colors, at least temporarily. Within the silver family, there are warm and cool versions, mouldings with different tones of antiquing glaze over the silver, as well as shiny and matte finishes to choose from. Stained wood mouldings all tend to have warm finishes, but the range of colors is still broad. You can make your choice easier by narrowing down to a specific range of color. You may think black is black, but even that can vary. Some blacks are darker than others. Or like silver, they can be warm or cool.

Step by step, you can broadly eliminate options until you are left with the best one for your project. If you find that the color you thought would work best isn't working so well after all, then it's time to re-focus, choosing a new color and starting the process again.

A well-trained frame designer is not overwhelmed by the options and has the ability to compare various colors at one time. If you are able to do it, by all means do. If not, stick to one color at a time to make things easier. This process should be fun, not overwhelming.

Using this tone-on-tone frame, mat, and fillet combination helps make the art more prominent.
("It Once Was Home II" by Greg Perkins. Courtesy of Artaissance, www.artthatfits.com)

This frame has a muted pink panel, which may seem odd with the autumnal colors in the art. When you begin to study the art, you see a pink/purple cast coming through all the other colors, and it all makes sense.
("Spring Lake" by Camille Przewodek. Courtesy of Drybrush Graphics)

GUIDELINES FOR SELECTING A MOULDING COLOR

• Relate it to the subject of the art

Typically, stained wood frames look good with outdoor scenes and painted finishes are great for colorful children's art.

• Consider the age or era of the art

Colorful frames were not used in the late 1800s so it is best to avoid them when framing something from that period if you want the results to look authentic to that time.

• Consider the mood of the art

A photo of a landscape on a stormy day may look best with a subdued gray frame, but a photo of the same scene on a sunny day may look better with a gold frame.

• Avoid frame colors that are lighter, brighter, or darker than the colors in the art

Strong color options cause the viewer to notice the frame more than the art. When frame colors are too dark, the art can look washed out, as if it is faded. When the colors are too light, they make art look even darker. When framing colors that are brighter than the art, they can make the art look lifeless.

• Look at the amount of contrast in the art

The stronger the contrast is in the art the more likely it is that you can choose a frame with more contrast. When the art has low contrast and the framing is strong, it reduces whatever visual strength the art might have.

• Although the primary goal is to enhance the art, be aware of how the frame will look in the room

If the framing doesn't look good with the decor, it won't be flattering to the art either. A lot of people try too hard to match their framing and art to the room where it will be displayed. When it does match, it blends right in and sort of fades away. Instead, try creating interesting accents that will call more attention to the art.

• Neutral framing designed to look good with the art is easier to move to a new decor or to your next home

A quality piece of framed art can last for decades or generations. If the frame design has a degree of versatility, it can be highly effective in a variety of situations.

• Determine if the color scheme of the art is mostly warm or cool and frame accordingly.

These silver spoons are neutral, leaving many options for the moulding color choice. This frame is silver, relating to the spoons, with a blue panel. The blue could be a way to coordinate with a specific room.

The color of this moulding blends with the matting, maintaining an earthy color scheme. The carved wood artifact is much lighter and therefore remains prominent.

This frame is red, but it is a subdued tone. It is perfectly fine to use color but, when doing so, make sure the art remains the strong focal point. ("Venice, Palazzo" by Elizabeth Ochwell. Artist represented by www.islandinternational.com)

Vibrantly colored artwork can be fun and exciting to frame. The bold colors and high contrast of the art allow for more choices than with low contrast art.
("Skribl #1" by Campbell Laird. Courtesy of Artaissance, www.artthatfits.com)

This piece of art has moody quality. A dark frame helps to maintain that look. A subtle red wood frame serves as a liner, adding depth and interest to the design.
("The Good Life" by Jane Bellows. Courtesy of www.newerahd.com)

The same piece of art framed in a much lighter design. It looks nice but loses some of the mood, and the light mat commands a lot of attention.

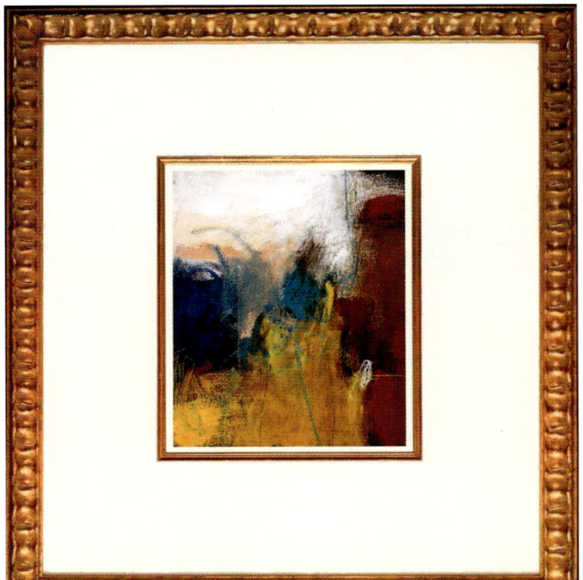

This artwork contains both warm and cool colors so it works with either gold or silver frames. Consider how each choice will work in the room where it will hang.
("Lagoon" by Charlotte Foust. Courtesy of Artaissance, www.artthatfits.com)

When framing art, consider the smallest details. The brilliance of the gold here is beautiful with the art but may not work in a room with softer finishes.

By comparison, the gold frame at right is muted. Although the frame is larger in scale, the subtle gold finish could work in a space where the other gold frame does not.
("What a Pair" by Robert Perkins. Private Collection of Greg Perkins)

MOULDING STYLE

Frame mouldings can be selected in several different ways. Some framers strive to use the historically correct style for the art. I am not a purist, so I am more apt to mix art and frame styles. If there is a newer moulding design that looks beautiful with the art, I may actually favor it, knowing it will bring a fresh perspective to the art and therefore to my frame designs. I can't help but think that if the artists and framemakers from past centuries were with us today and could see all of the great moulding choices we have, they might enjoy trying something different, too. With that said, I do see a time and place for sticking with tradition.

When pieces of high quality antique art or important historical items are being framed, authentic frame styles make a lot of sense. There are a number of finished corner frames and custom framing options available for this. On the other hand, a piece of new art painted in a historical style offers an opportunity to use a beautiful new frame style. Change can be a good thing. Just use good judgment as to when it is important to choose the proper style and when it may be desirable to try something new.

A Concerto finished corner frame is the ultimate surround for a piece of art. The bold colors and strong outlines in this art stand up to the dramatic design of this Spanish-style frame. Together, the art and frame make a grand statement.
("Room Interior" original pastel by Ruth Franklin. Courtesy of VINSONart.com)

This Concerto finished corner frame is an Italian style with sgrafitto designs scratched into the surface. Even though the art is entirely different in style, the art and frame both have intricate details that help them work together.
("How Now" by TL Lange. Courtesy of Artaissance, www.artthatfits.com)

There is a primitive quality that shines through the contemporary style of this art. A traditional moulding with a tribal pattern tones down the more modern elements. This design may be the perfect choice if the art is being framed for a traditional or culturally styled room decor.

Look at the difference when a simpler frame surrounds the same piece of art. The primitive aspects are played down and the contemporary styling comes out.
("Joy Series" by Edosa Oguigo. Courtesy of Artaissance, www.artthatfits.com)

GUIDELINES FOR SELECTING MOULDING STYLES

• *For historical pieces, choose a frame style that would have been used at the time the art was created or relates to the era that the art depicts*

This is a good way to give the impression the framing is original to the art.

• *Choose a frame reflecting the specific style of art*

There are mouldings that are designed to reflect the Art Deco and Art Nouveau periods or the Southwest and Cottage styles, to name a few.

• *Choose a frame inspired by the same country as the art*

For example, a frame with delicate French ornamentation can be used on a piece of French Impressionism with fine brush-strokes. Spanish frames are usually bold in style and ornamentation. They work well with Spanish art, such as prints by Picasso and El Greco.

• *Study the shapes in the art and choose a moulding with similar shapes*

A square profile can be used for art with geometric shapes, or a rounded profile will relate to a picture of daisies having rounded petals. Also look to the shapes within the ornamentation to coordinate with shapes in the art.

• *The most important thing is to choose a frame to enhance the art while also considering the room decor*

If you have several frame design options you like equally, but one also suits the room decor better than the others, by all means choose it. Just don't force the room style onto the art when it isn't appropriate.

(Top left) The climbing roses in the print were the inspiration for this moulding choice. The flowers look more like mounds of color than individual flowers. The bold shapes of the frame's ornamentation have a similar appearance, providing a stronger relationship than a frame with a delicate pattern.
("Full Bloom" by Elena Ilku, Courtesy of Artaissance, www.artthatfits.com)

(Top right) This Southwest Style carved moulding is a beautiful complement to the artwork. Although the specific pattern on the frame isn't in the art, the two work well together because of their common style
("Taos Series, Ed at Ranchos", by T.C. Lillick. Courtesy of Artaissance, www.artthatfits.com)

(Left) Even small details make a difference. Notice how the small curving lines on the frame's ornamentation mimic the antennae of the butterfly.

When different framing options work equally well with the art, let the decor of the room be your guide, helping you make an appropriate choice.
("Morning Walk" by Ted Goershner. Courtesy of Artaissance, www.artthatfits.com)

(Above) Look closely at the pattern on the moulding and the circular pattern at the top center of the art. While they are not identical, they play off each other to create a custom design.
("Untitled" by TL Lange. Courtesy of Artaissance, www.artthatfits.com)

(Above right) This print fits in beautifully in a cottage-style decor. The embossed white moulding also suits that style, providing a common theme to unify the presentation.
("Big White" by Soroya Chemaly. Courtesy of Artaissance, www.artthatfits.com)

(Right) A single piece of art can often be framed in very different ways. Sometimes the unexpected way provides a lot of interest. In this case, the contemporary art is surrounded by a traditional frame. The color serves as one bond between the two.
("Untitled" by Marko. Courtesy of Progressive Fine Art)

When the center of interest in the art fills the paper or canvas, the resulting shape generally requires a substantial frame to look supportive of the perceived visual weight. Here an opulent Spanish-style frame (left), combined with a silk liner and an unexpected red fillet provide a 7" wide frame that is well-proportioned to the figure. (Right) A narrower moulding leaves the art hungry for more. The scale of the frame might work if a substantial mat border was also used, but the frame alone lacks the necessary power to balance with the art.
("Intrinsic Thought" by Jinous Kani. Copyright Deljou Art Group)

MOULDING SCALE (WIDTH AND DEPTH)

Once the desired frame color and style are established, selecting the best moulding scale comes next. This can be tricky, both to do and to explain. Some small pictures look best with frames made from wide mouldings, and some large paintings look good with narrower frames. The best moulding width and depth should be based on the individual qualities of each picture. For any piece of art, the appropriate moulding scale may also vary depending on the scale of the area where it will be displayed. And if the art will be a part of a grouping, it may not need as much framing as it would if the same pieces were displayed individually.

Moulding width is an important consideration in framing because an appropriate width helps create a sense of balance between the frame, the art, and all the other framing components. It also helps the completed piece look its best in the room where it will be on display.

Moulding depth may not seem especially important, but it is something that needs to be considered. You should always be conscious of the depth if you want a frame to be capable of holding everything that will need to go into it, including glass and backing boards. Choosing a deeper frame moulding than you actually need also allows you to project the art forward for easier viewing or to recess it into the frame to enhance the perspective or depth that is inherent in the art.

It may seem appropriate to use narrow moulding when framing small works of art, but a wider moulding like this adds significance that will help the art receive attention that it may not otherwise get.
("De La Mundo II" by Greg Perkins. Courtesy of Artaisssance, www.artthatfits.com)

When framing a canvas, consider wider mouldings. Keep in mind that the width of the frame can make up for the lack of matting.
("Veilied Valley" by David Marty. Courtesy of Artaissance, www.artthatfits.com)

GUIDELINES FOR DETERMINING MOULDING WIDTH

• *Consider the overall size of the art*

When framing a large and a small version of the same image, the larger one will need a wider frame to maintain the same degree of balance that a narrower moulding brings to a smaller print size.

• *Look at the size of the focal point and other shapes within the art*

Two pictures that are the same size may require two different frame widths. For example, a close-up portrait may need a wider moulding than a picture of a person in the distance. This is necessary to balance the moulding with the larger, bolder features in the face.

• *Relate the moulding size to the visual weight of the art*

Dark colors, heavy textures, opaque paints, and big shapes all increase the perceived visual weight of art. The more visual weight something appears to have, the wider the frame will probably need to be to look visually supportive.

• *Consider the color contrast between the frame and the art*

Imagine a lightly colored piece of art. If you choose a darker-colored frame, it should be quite narrow since there isn't much visual weight to the art. A lighter frame could be wider because it will be in balance with the art.

• *Relate to the era of the art*

For example, wide mouldings were popular in Victorian times. Using wider frames when framing art of that period or of that style can lend a look of authenticity to your frame design.

• *Use wider mouldings to help call attention to small art or objects*

When choosing a frame, be sure to consider the era of the art. If your goal is to create a look reminiscent of that era, the width of the moulding is an important part of the equation. This Old World-style still life looks right at home with a moulding as wide as which the art may have originally been framed in.
("Seville Orange" by Jean-Baptiste Simeon Chardin)

GUIDELINES FOR DETERMINING MOULDING DEPTH

• *Choose a frame that is deep enough*

Think about everything that has to fit into a frame. When you see an unframed print, remember that glass, mats, backing boards, and other frame elements may also have to fit inside it.

• *Use deep mouldings to project the art*

Many people only think to use a deep moulding when framing a dimensional object that necessitates it. Depth can also be used on two-dimensional images like photos or prints. If the art is placed forward in a frame, it will project towards the viewer. This can be helpful if framing is displayed in a bookcase.

• *Use a variety of moulding depths for a grouping*

If you are going to frame a variety of art or photos to create a grouping, choose varying moulding depths to add interest on the wall.

• *Recess two-dimensional art into a deep moulding*

If a print has inherent depth, such as the perspective from looking up a city street or down a garden path, you can enhance the depth by placing the picture halfway back into a deeper moulding. Avoid placing it all the way back as it may get lost in the shadows.

LINERS

The name "liner" generally refers to a fabric-covered moulding. A liner is used inside another frame, most often when framing canvases. A liner provides visual relief between the art, the frame, and the surrounding room decor.

Some liners are not fabric-covered. Another moulding can be used, including those with painted finishes for a splash of color. Or a plain gold moulding might be used inside an ornate gold frame to provide visual relief without the distraction of another color. Simple liners are great to use when a wide, dark, or heavily ornamented moulding seems to crowd the art. A liner can visually separate a heavy frame and its pattern from the art, giving it breathing space to help the design work better.

Ornate frames can also be effective as liners, especially when used to surround art with simple backgrounds. Imagine a complex still life centered on a plain background. An ornamented liner may help the overall frame design relate better to the patterns and textures of the objects in the art.

(Left) You can make art on paper look like a painting on canvas. A double liner and deeply carved pattern on the frame add dimension and interest. (Right) In this design the piece is matted. The frame is a similar color to the one on the left but the mat is much darker, bringing an altogether different look to the art. *("Woman" by Yunessi. Courtesy of Deljou Art Group)*

(Above left) The pattern on this narrow liner is small in scale so it adds visual interest without overpowering the art. The liner is effective in preventing the wider, heavier frame from confining the artwork.
("Floral Botanical" by Sue Gillette, Courtesy of Artaissance, www.artthatfits.com)

(Above right) Liners typically look best when they have a different width than the frame. This narrow liner, which has a lighter color than the wider frame, helps draw attention back to the object.
(Architectural Ornament. Private Collection of Greg Perkins)

(Right) Patterns do not always have to relate directly to the art. In this case the minimal pattern on the liner adds interest without detracting from the art.
("#5 (From This Point On)" by Rodney White. Courtesy of Artaissance, www.artthatfits.com)

GUIDELINES FOR CHOOSING LINERS

• Choose a liner that lets the central point of interest in the art stand out in the completed frame design

• The width of the liner should differ from the frame surrounding it to avoid a design that looks static and uninteresting

• Liners serve as a transition between the details of the art and the frame

A liner may need to be plain or ornamented to make that transition work.

• The texture of a liner can coordinate or contrast with the item that it surrounds

For example, a linen liner continues the texture of a canvas, whereas a smooth, painted wood liner contrasts with it. Either can look appropriate.

• Combine liners to add depth and interest

• Use other mouldings as liners when you want a different look

The plain gold liner used inside this frame is called a slip. It is a nice alternative when you want to add visual relief in a more minimal way than using a dimensional liner.
("Still Life w/ Pears" by Mary Mackey. Courtesy of Deljou Art Group)

Historically, landscape paintings were framed in gold. Over the years framing for this type of image has become more casual, but there is nothing like a fine water-gilded frame to set off a traditional plein air *(outdoor) painting.*
("Cottonwood II" by Kim Coulter. Courtesy of www.newerahd.com)

This liner is a painted wood frame. The simplicity of this moulding helps offset the intricate pattern on the wider frame.
("Chrysanthemum I" by Filipo Ioco. Courtesy of Grand Image, Ltd)

In this design a white liner coordinates with the fern. The liner minimizes the amount of white so it doesn't overpower the art the way a wider white mat might.
("Fern Plate No. 714" Original artwork copyright D'Schommer. Print copyright Wild Apple Graphics)

Rather than using a linen-covered liner, the sleeker finish of this liner allows the textured look of the canvas and artist's brushstrokes to stand out beautifully.
("Sonnet" by Maeve Harris. Courtesy of Grand Image, Ltd)

FLOAT FRAMES

A float is most commonly used when framing art on canvas. A standard frame moulding has a lip that holds the glass and other contents in place. Float frames are different in that the art drops in from the front, allowing the entire front of the canvas to show. They are now available in much more interesting profiles than in the past. Float frames can fit snugly around the art, but it is more common to back them up a small distance, providing breathing space for the art. If an artist continued a painting down the sides of the canvas, it is desirable—if not necessary—to show it, so the space helps.

Sometimes float frames are used as the only frame. This creates a very sparse, minimal look that may or may not do justice to the art. A float frame should be looked at as a method of presenting a painting so that none of it is hidden under the frame. If that presentation isn't enough, add to it.

If you choose to add more to your design, a standard frame moulding, a liner inside a moulding, or a stacked moulding combination can be added around it to bring out the specific personality of the art if the float alone doesn't do the job.

GUIDELINES FOR USING FLOATS

- ***Use a float frame when a painting has nicely finished edges that are desirable to show***

If an artist painted an important feature right to the edge of the canvas or perhaps even continued the painting down the side of the stretcher, a float prevents those details from being covered

- ***Keep both the width and depth of the float in mind so you know how it will fit the canvas as well as any frame you use to surround it***
- ***When appropriate, add a frame around the float to complete the design by enhancing the color and style of the art***

The float frame used around this painting allows the entire surface of the art to show. However, the float didn't do anything to bring out the personality of the art. The wider gold moulding, with its stepped back profile shape, relates to the geometric background that the artist painted and provides a more finished look.
("Landscape" by Jeff Lange. Mixed media landscape from www.langeart.com)

This print on canvas is framed using just a float moulding. A float provides a very minimal edge that works best with streamlined, contemporary artwork.
("Frill C" by JB Hall. Courtesy of Artaissance, www.artthatfits.com)

When a fillet is added inside a mat opening, it provides an outline that helps to draw attention from the frame back to the art.
("Evening Glow I" by Kim Coulter. Courtesy of New Era Portfolio, www.newerahd.com)

Without the fillet, the ornamentation on the frame commands more attention and is therefore distracting from the art.

When framing dark pieces of art, it may be tempting to try to lighten them by choosing lighter mats and frames. If you do this, the strong contrast tends to make the art look even darker. Rather, use dark mats and frames with small details in the lighter color as shown here.

FILLETS

At a glance, a fillet may look like a small frame. Upon closer inspection you will see that it doesn't have a lip like the one that a frame uses to hold such contents as art, glass, and mats. Most framers call them "*fill*-ettes," although some call them "fill-*ays*."

A fillet's purpose is strictly for use as an accent. Fillets are generally used inside the opening of a mat, but they also look great inside the lip of some frames. Fillets are often manufactured along with collections of frame mouldings, so they come with matching finishes as well as coordinating textures or patterns.

A fillet brings an added look of quality to a frame design. Since it has dimension and perhaps an interesting shape, it also adds another layer of visual interest to custom framing. Fillets have been used for many years. They have never gone out of style, so they offer safe, classic details that won't soon become dated.

Fillets made from wood are acidic, which means they can potentially harm an item being framed. There are products and techniques framers can use to prevent this, such as sealing the exposed raw wood and placing a mat between the art and fillet to block direct contact.

Fillets are typically used inside the openings of mats, but they also make a nice addition sitting in the lip of a frame, as shown here.
("Essence of Life" by Heidi Coutu. Courtesy of www.editionslimited.com)

GUIDELINES FOR DESIGNING WITH FILLETS

- ***Select a fillet that matches the moulding color and place it inside the mat opening to create a cohesive and classic frame design***
- ***Use a contrasting fillet inside the frame lip to add a new look to the frame's inherent style***
- ***Use two of the same fillet, repeating it inside the mat opening and the frame lip to pull the entire design together***
- ***In general, use wider fillets with wider frames and narrow fillets with narrow frames for balance***

A wider fillet has a greater ability to pull attention back to the art from a wide or heavily ornamented frame.

- ***Stack fillets***

When a matching fillet color isn't available with enough width to balance the frame, use two of the same fillet designs side by side to create the necessary scale. Use two different fillets side by side for added interest, such black with gold or a silver fillet with ornamentation alongside a plain silver fillet.

You can use two contrasting fillets to add depth inside a single mat. The fillets also create a strong band around the art to draw the viewer's attention back from the frame to the art.
("Regal Dragonfly III". Courtesy of World Art Group)

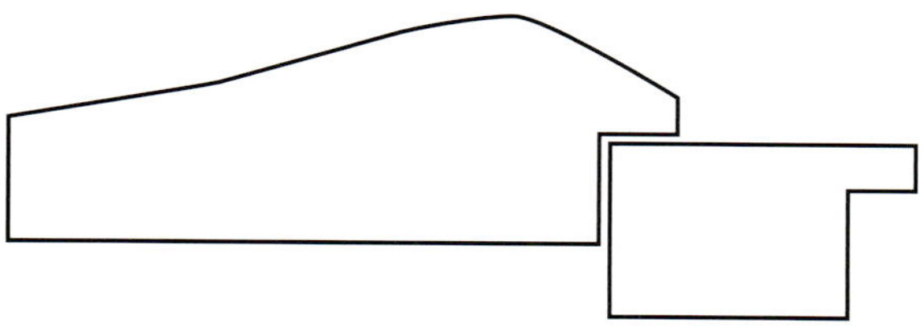

The two mouldings used for this frame combine both the square and angular line work in the art. When you can find common ground between the shapes in the art and frame, it helps to unify the presentation.

("Welcome" by Jaime Ellsworth. Courtesy of Artaissance, www.artthatfits.com)

ENHANCERS

Enhancers are also similar to small frame mouldings. Like fillets, they are not intended to be used as frames. They come in several colors and a variety of patterns. They are designed to add definition to a frame while also possibly altering its style. For example, using three different Enhancers can completely transform the style of a frame. One of these Enhancers may give the frame design an Asian flair, while the other two may make it look nautical or elegant.

Here an Enhancer was used to add definition and interest to the frame design.
("Alphabet Soup" by Greg Perkins. Courtesy of Artaissance, www.artthatfits.com)

The pyramidal shapes on this Enhancer relate to the geometric shapes in the art.
("Will He Ever Make a Move?" by Greg Perkins. Courtesy of Artaissance, www.artthatfits.com)

This is a 30"x30" canvas with large details in the image. To relate to that scale, two mouldings were combined. Since they both have the same finish, they appear like one wider frame moulding.
("Wine for One" by Ray Pelley. Courtesy of Artaissance, www.artthatfits.com)

The copper tone of the inner frame doesn't just relate to the colors in the art. It also coordinates with the undertones on the rubbed areas of the predominant outer frame, giving the entire design a unified look.
("Warm Afternoon" by A Pell. Courtesy of Artaissance, www.artthatfits.com)

Unusual combinations of mouldings can make some of the most interesting frame designs. This pastel burl wood moulding is combined with a copper finish frame. Each one ties to colors in the art. In addition, the way the satin finish of one moulding contrasts with the metallic sheen of the other further enhances the design.
("High Rise Circus" by Scott Hile. Courtesy of Rosenbaum Fine Art)

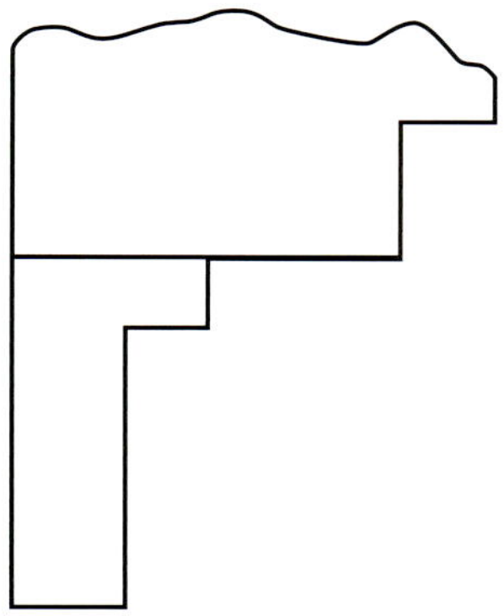

When framing an object, always be aware of the depth you will need, not just for the object itself but also for the glass, backing, mats, etc. You can expand your shadowbox options by combining mouldings. The frame you see in this photo is actually quite shallow but a second, deeper profile was placed behind it to build the necessary depth to hold all of the contents.

STACKING MOULDING

Stacking is a technique of combining two or more mouldings into a single frame. This practice has been used for centuries. There are a variety of reasons stacking can benefit a frame design.

REASONS TO DESIGN WITH A STACKED MOULDING COMBINATION

- Stacking one moulding beside another will add width to create a more substantial frame.
- Stacking one moulding behind another adds depth. This can help when the art and all of the components won't fit into a specific frame you want to use.
- Insert a small, colored frame inside a gold or wood tone moulding to add an accent to the frame design that accentuates color from the art.
- Combine diverse frame styles, such as rustic wood and ornate, to create a unique eclectic style.
- Stacking different moulding profile shapes together can result in a completely new and different look that may be more appropriate for the art.
- If you want to use a very narrow wood frame on a large or heavy picture, you can add a narrow metal frame inside it to support the weight. The wood frame can completely hide the metal moulding.

CHAPTER FIVE

MATTING

A mat is used either as a border between the art and frame or as a background to set art or an object on. It serves three distinct purposes:

• As a border, it lifts the glass off the surface of the art. This provides an air space so that condensation can't become trapped against the face of the art.

• It provides an area of visual relief between the art and the surrounding room, making it easier to focus on the art.

• It can add color as well as texture or pattern to a frame design. Keep in mind that its purpose is to enhance the art, not to become the focal point.

When mats are used as borders surrounding art, they may be just a single layer, but they can also have two, three, or even more layers. They can be cut with just one window opening for a single picture or with multiple windows to combine a number of pictures into one frame. Mat openings are most often square or rectangular, but ovals and other specialty cuts can be used. Mats should be understated enough so that they won't overpower the art.

Mats are also used as solid backings when placing objects in shadowboxes. To finish a shadowbox, strips of the same mat are typically used to line the inside walls of the frame to hold the glass and backing apart, creating space for the object(s).

When should a mat be used?

• To frame almost all paper art and documents.

• To serve as a spacer separating the glass or acrylic from the item being framed, although other types of spacers are available.

• To increase the size of what you are framing.

• To combine more than one item into a single frame. Multiple openings can be cut so that any number of things can be grouped into one frame.

• To provide a background to set or mount something on.

The circular openings in these mats relate to the rounded shape of the mask. When you blend the mat opening shape with the shape of the object, it helps keep attention focused on the object. A square mat opening would conflict with the mask and leave the viewer wondering which to look at.

(Mask by Larry Laslo. Courtesy of Rosenbaum Fine Art)

Multiple openings in mats allow you to combine more than one piece of art into a single frame. When doing this, use narrower bands between the images and wider borders around them.
("3 Antique Bookplates". Private Collection)

Blending the mat color to the background of the photo helps put a strong focus on the person. An oval mat opening was used to eliminate a lot of the background, focusing on the portrait.
(Photo of the Author's Mother)

Specially cut mats can be a great thing if and when they make sense with the art. In this case, the square opening in the mat was cut at a diagonal, creating a more dynamic presentation. A vertically elongated mat also adds interest.
("Life Stories" a detail by Greg Perkins. Courtesy of Artaissance, www.artthatfits.com)

A multiple opening mat is a great way to present a set of prints. You have a choice of a vertical or horizontal presentation. Consider the space where the completed piece will hang to determine which orientation may be the best.
("Fragments I, II, and III" by Greg Perkins.)

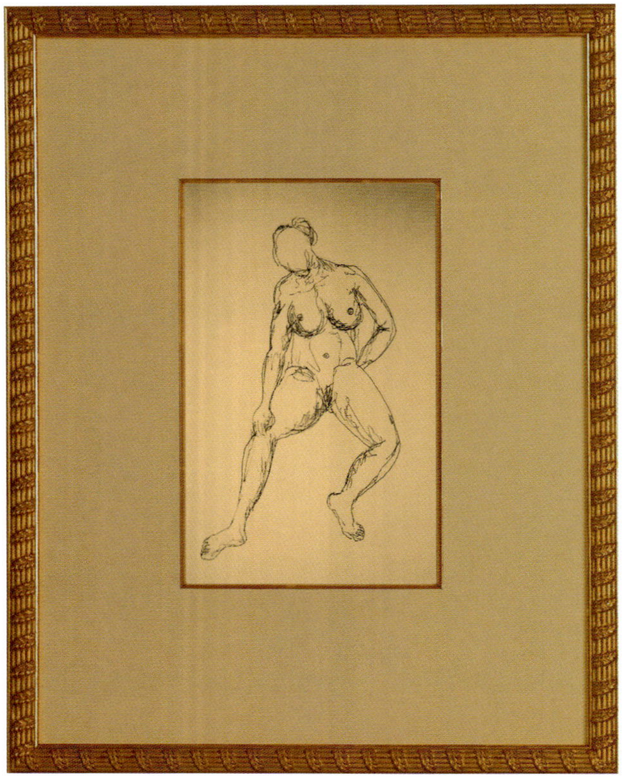

This pen-and-ink sketch has become discolored over its life due to the poor quality of paper it was drawn on. When it was reframed, a darker mat was selected, mimicking the age spots on the art and leaving the remainder of the art looking fresher.
("Figure in Motion" by Greg Perkins)

Although you may be attracted to colorful mats like this one, it is common to tire of bold mat colors faster than those that are neutral. A neutral color also allows you to easily move the art to a new home or decor.
("Moth Orchids" by Elizabeth Horning. Courtesy of Artaissance, www.artthatfits.com)

MAT COLORS

When mats were first introduced, they were all neutral colors like off-white, cream, and pearl gray. Mats are now available in hundreds if not thousands of colors. Sometimes, having so many options adds confusion to the decision-making process. On the other hand, having all these choices means being able to match almost any color you want. Even with all the color choices, neutrals remain the best sellers. Sticking with neutrals adds to the longevity of your frame designs. The most colorful mats are usually only added as accents under a predominant mat. However, there are times when more color is needed and all these choices are welcome.

GUIDELINES FOR CHOOSING MAT COLORS

- ***Relate to the era of the art***

Subdued colors are appropriate for Old World artwork. Mid-Century Modern, Art Deco, and other styles all have their own ranges of colors you can use, too.

- ***Relate to the subject of the art***

Use bright colors for a happy clown print. More neutral colors will look better with Dad's business portrait.

- ***Use neutral mat colors to create classic frame designs***

Popular colors schemes go out of style, but neutrals last.

- ***Neutral matting makes it easier to move framing into a new home***

You can extend the lifetime of your investment with neutral mats.

- ***Avoid mat colors that are lighter, brighter, or darker than the colors in the art***

The art should remain the focal point.

- ***Blend the mat to the background colors in the art***

More often than not, the most detail and interest in the art is near the foreground. By coordinating mat color with the background, those details will stand out.

Many prints can be matted with either light or dark colors and look equally good. When both choices will work, pick the one that will coordinate best with your decor.
("Pink I" by Sang Duk Park. Courtesy of Artaissance, www.artthatfits.com)

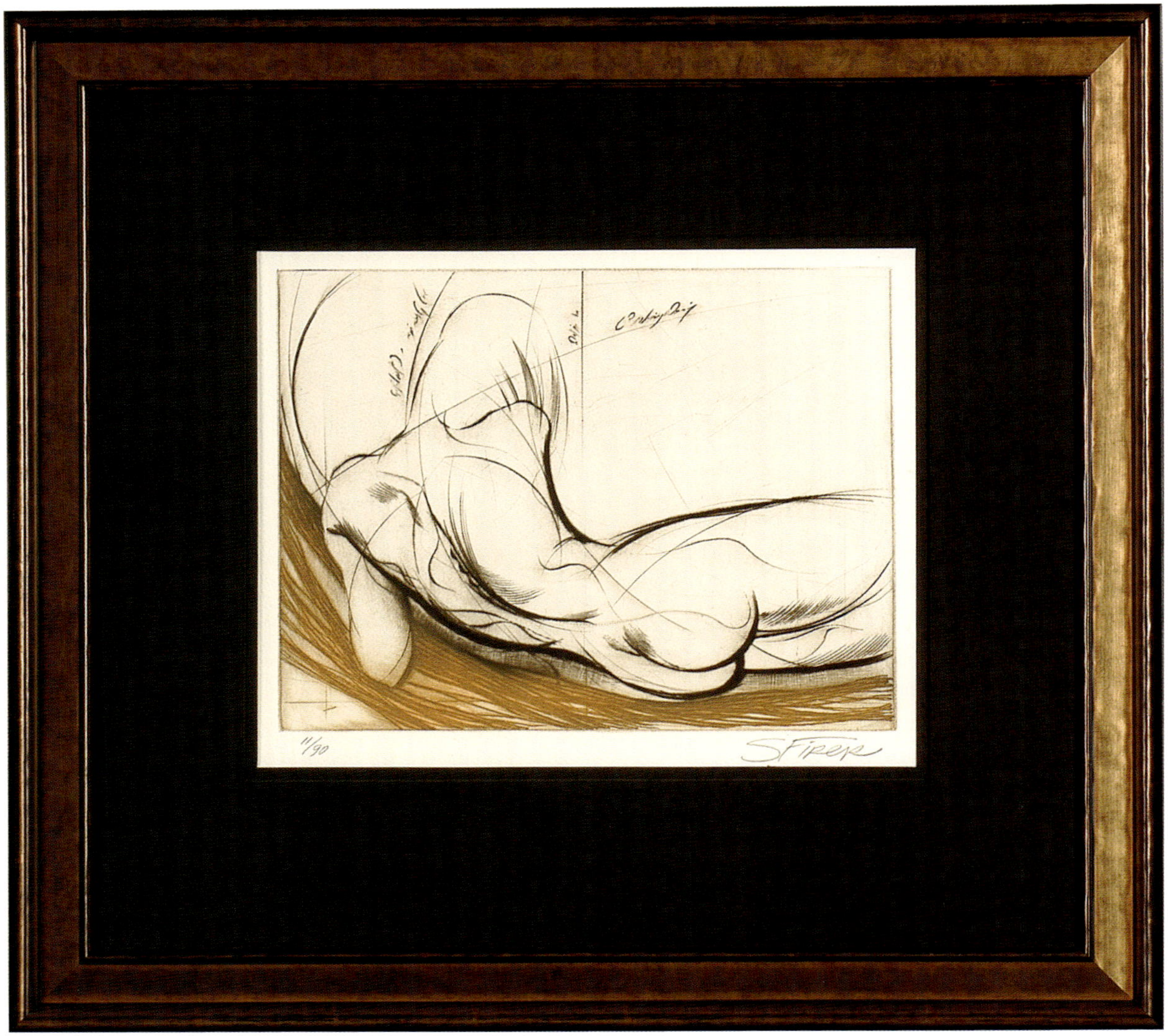

Because the subject in these two pieces of art are figures, it's important for them to stand out in the completed framing design. The black mat adds drama but the white mat with black bevels helps call attention to the figure.
("Focus IV" & "Focus VI" by S. Firer. Courtesy of Progressive Fine Art)

When a mat color other than neutral is desired, look for a tone like this subdued green that allows the colors in the art to remain prominent

Hints of brown peek through many of the other colors in this artwork, creating an overall pleasant blend between the mat and image.

If you prefer a light neutral mat, this choice works because it is slightly darker than the white clouds and has a yellow tint that relates to other colors in the art.

This mat combines the three colors shown independently in the other examples. This is a great choice because it has a neutral appearance but still provides a sense of color.

("Passing Storm" by David Marty. Courtesy of Artaissance, www.artthatfits.com)

• Use dark mats for dark art and light mats for light art

The strength in dark pieces of art usually comes from highlights that are light in color. A light mat competes with those highlights, whereas a dark mat allows them to shine. The opposite is true for light artwork.

• Determine the approximate percentage of each color in the art

Generally, any of the two or three most predominant colors will look good as a single mat or in combination as a double or triple mat.

• It isn't always necessary to match the mat to the colors in the art

For example, a khaki-colored mat may coordinate with the greens and browns of a landscape and look great with it even if that exact color doesn't appear in the art.

• You don't always need multiple colors for a multiple mat

You can use two or three different colors for a double or triple mat, but it may look nice to use just one color.

• Use neutral mats to elevate the perceived value of the art

Since many people are accustomed to seeing white mats used with fine art in museums and upscale galleries, such mats sometimes give the impression that the art is better than it is.

Traditional landscapes generally look best matted with earthy neutrals such as the light gray-brown or dusty green shown here.
("Foggy Day," by A. Pell. Courtesy of Artaissance, www.artthatfits.com)

(Top left) To keep the focus on the artwork, avoid mat colors like this bright white that are lighter than the lightest color in the art.

(Bottom left) Also avoid mats that are darker than the darkest colors in the art. In this example, the black mat is so deep and rich that it overpowers the subtleties of the art.

(Top right) This midtone tan mat is softer, keeping the focus on the art.

(Bottom right) If a darker color is favored, this one is good because the darker color in the art remains strong.

("Pompelmus Copper" from New York Botanical Gardens. Courtesy of Artaissance, www.artthatfits.com)

The aged color of this patterned mat works well with the timeworn look of the tobacco tins being framed.

A good rule of thumb is to use light mats on art that is predominantly light in color. It helps maintain the color balance present in the art.
("Parrot Tulip I" by Artin. Copyright 2003, Tom Artin; published by the McGaw Group, www.bmcgaw.com)

One method for selecting mat colors is to blend the mat to the background color of the art so the focal point in the foreground stands out. However, in this case, the background is such an intense color, that the mat pulls a lot of attention away from the art. This particular color may also be challenging to coordinate with the decor in many homes. A more neutral mat selection would have been a better solution.
("Sunflowers" by Van Gogh)

When the art is dark, stay with dark mat colors. Whatever light details there are will stand out versus using a light mat that competes with them.
("Double Parrot Tulip" by Prince. Copyright 1997, Len Prince; published by Bruce McGaw Graphics, www.bmcgaw.com)

(Top left) A dark mat blends with the darker tones in the art, making the light colors stand out and become prominent.

(Top right) A light mat competes with the light features in the art, diminishing their importance and making the already dark colors in the art appear even darker.

(Bottom right) A mid-tone mat color allows both the light and dark colors to show up equally well.
("Exotic Lady III" by Marta Wiley. Courtesy of Artaissance, www.artthatfits.com)

The color of this mat doesn't actually appear in the art. It still works because it is a cross between several of the colors that the artist used.
("Bales in the Back 40" by Christa Malay. Artist represented by www.islandinternational.com)

When using a double or triple mat, it is common to choose different colors for each layer. A sophisticated alternative is to repeat the same color for all layers.
("Abraham's View" by Danna Harvey. Courtesy of Artaissance, www.artthatfits.com)

This mat is quite light for the rich colors in the artwork. Typically, it would be best to choose a somewhat darker color to avoid calling so much attention to the mat. However, white rag mats are quite standard in museums and upscale art galleries so people are accustomed to seeing them. That makes it quite acceptable to use them.
("Room Interior" Original pastel by Ruth Franklin. Courtesy of VINSONart.com)

A neutral top mat was used here, but two accent colors were selected, based on their dominance in the art. Because there is more yellow than red, more of the yellow mat shows.
("Gilded Tile II" by Angeles Maceira. Courtesy of www.editionslimited.com)

e light pastel colors of this map could easily become verpowered by a darker or ore colorful mat. By keep-ing the color scheme light and airy, the few darker details in the map show up the best that they can.

The neutrality of this earthtone mat helps it blend with other similar colors in the art so the colorful trees stand out.
("Row of Trees II" Original art by Norman Wyatt, Jr., Print copyright Wild Apple Graphics)

Some people would choose a red mat for this art to bring out the color of the birds. In reality, the red competes with them rather than enhancing them.

A light neutral mat is a better choice if you want the red birds to remain dominant. You could add a narrow red accent if you want some color.
("Red Birds" by Ososki. Courtesy of www.editionslimited.com)

Mats and fillets will enhance objects just as they enhance artwork. The black mat is a woven texture that somewhat mimics the black needlepoint purse. The mat behind the purse is suede so the texture of the needle-point becomes more prominent.

The fuzziness of a suede mat has a comparable look to the sand, relating to the beach theme of the art.
("On the Beach" by Winslow Homer.)

This art photo has a hazy quality that lends itself to a softer looking mat. Suede was a good choice to tie to the look.
("Mystic III" by Pezhman. Copyright Deljou Art Group)

This clay and wire sculpture could hang on a wall unframed. By mounting it to a contrasting mat and framing it, the mask suddenly looks more important and will show up well on nearly any color wall.
("Ibu" by Greg Perkins. Private Collection of Greg Perkins)

PATTERN/TEXTURE

Some mats come with textured or patterned surfaces. These are made with several types of fabrics that are commonly applied to mats. They include linen, silk, and suede. Sometimes simulated versions of fabrics are also printed onto mats. A host of other patterns are also available, and custom framers can attach other fabrics or papers to matboard.

GUIDELINES FOR USING PATTERNS AND TEXTURES

• ***Avoid any pattern or texture that will detract from the art***

• ***Use patterns or textures only when they make sense and relate to something in the art***

Patterns or textures should enhance the appearance or the perceived value of the art.

• ***Use a contrasting fabric behind objects***

For example, a silk mat will set off a velvet purse whereas a suede mat will blend with it so that it won't show up as well.

• ***Do not choose patterns or textures that are larger in scale than a similar look in the art***

Using mats with reduced-scale versions of the patterns or textures in art can create a custom look without detracting from it

• ***Relate to the subject of the art***

A coarse linen mat might be right for a rugged landscape while silk with a delicate weave may fit with a bridal portrait.

• ***Relate to the era of the art***

For example, since velvet was often used in Victorian times, you can use a suede mat to simulate that look when framing art from that era.

• ***Use bolder mat patterns and textures as accents rather than as primary mats***

It is easier for most people to live with softer colors and patterns for longer periods of time. Strong mats can either date a frame design or simply wear on you.

(Top) This mat has a very minimal flannel pattern that softens the mat surface without being too strong for the art.

(Bottom) This darker mottled pattern is also good with the art because it is nondescript and doesn't call much attention to itself while adding interest to the design.
("Woven Vessels II" by Albena Hristova. Courtesy of Wild Apple Graphics)

This mat is silk covered. That may seem like a strange choice for a picture of a rural house, but it works both in color and texture. Because the weave runs horizontally, it coordinates with the horizontal lines in the art.

("Rural Roses" by Thomas William Jones. Courtesy of Grand Image)

Fabric mats provide a great range of textures that work exceptionally well when framing objects. This linen mat provides a contrast to the felt, leather, and burl textures of the objects placed against it.

A black suede mat is much darker than other black mats. This print has a glossy finish so the black in it is very dark, making the suede a good choice. Since dogs are soft and fuzzy, the suede also has the right feel for the subject.

("Holiday by Rail" by Johanna Kreisel).

This mat has a very earthy quality that relates well with the art. It actually looks similar to the soil that the figures in the art are standing on. Even though the pattern isn't a replication of the art, it evokes the feel of it.

("Afrique". Private Collection)

This mat has a mottled, timeworn quality that works well with these vintage dried fern fronds. A crisp, fresh looking mat may look out of place.

(Top left) This design has a triple mat, with each layer having a different pattern or texture. A minimal linen-look is used as the predominant border.

(Top right) In this example a much bolder pattern appears in the dominant position. As a result, the mat commands too much attention.
("Summer Meadow" by April Richardson. Courtesy of Artaissance, www.artthatfits.com)

When you want a small picture to fill a larger wall space, use generous mat borders. In this case a triple mat was used. The black middle layer helps break up the wide border, making it look more proportional to the photo. Without that line, the mat may overwhelm the small photo.
("Tin Zebra" by Linda Omelinachuk. Copyright Linda Omelianchuk, www.omelianchuk.com)

Nearly anything can be framed to a different scale to help it balance with a large or small space where it will be displayed. The piece above right has a wider frame moulding, wider mat border, and an additional mat and fillet, enabling it to work well in a large space. For a smaller area, the simpler design on the right may be preferable.
("Autumn Warmth" by David Marty. Courtesy of Artaissance, www.artthatfits.com)

MAT BORDERS

The mat borders surrounding art can be any width so long as the outer dimensions are no bigger than a sheet of matboard, which is 32"x40", with some of the more popular colors are available in 40"x60".

Many frame designs could be improved if only they had more generous mat borders. When the borders aren't wide enough, the frame often appears to crowd the artwork, resulting in a frame design that looks "busy." Narrow borders can also degrade the perceived value of the art because they look skimpy. Obviously, too wide a border can be distracting, but a well-proportioned mat can help elevate the look and also provides the necessary visual relief to offset the art and help it look its best.

There are no rules for how wide mat borders should be for any given size of art. Some smaller pictures need wider borders more than larger pieces do. Different sizes of the same picture require mat borders of different widths. The shapes in larger images are bigger, so the mat borders should be wider to visually support them.

Typically, using a wider frame calls for wider mat borders. If a mat isn't substantial enough for a frame, the frame tends to confine the artwork. A more generous border will back the visual weight of the frame away from the art, giving it sufficient breathing space.

Small pieces of art can be framed with very generous mat borders. The key to doing this successfully lies largely in the relationship between the mat and frame proportions. A wider moulding is sometimes needed to create the right degree of balance with wider mat borders. A wider border may also need to be visually broken up by using multiple mats or by inserting a fillet an inch or so out from the art.

Generously proportioned mat borders lend a greater sense of importance to the pieces of art they surround. Wider borders also help fill a large wall.
("Orange Pop" by Greg Perkins. Courtesy of Artaissance, www.artthatfits.com)

When using wider frame mouldings on smaller pieces of art, be sure to increase the mat border widths to create symmetry or balance between them.
("Spring on a Slant" by Miguel Dominguez. Courtesy of Artaissance, www.artthatfits.com)

(Top) This is a powerful piece of art with imagery that fills the paper it is on. The art is floating on a background mat and then surrounded by a substantial mat border to allow breathing space between the art and frame.

(Bottom) In this design there is not a mat border. The frame comes in closer contact with the art so it doesn't get the visual relief like the design above.
("Buildings" Original pastel by Ruth Franklin. Courtesy of www.VINSONart.com)

UNIFORM BORDERS

When unframed art looks well balanced, the mat borders can be the same on all four sides and generally look fine.

GUIDELINES FOR UNIFORM MAT BORDER DIMENSIONS

• Avoid choosing mat borders the same width as the frame surrounding it

When the mat borders and frame are similar, they create a static look that can be uninteresting and prevents the eye from traveling effectively from the frame to the art. Usually, mat borders should be wider than the frame.

• When using wider frame mouldings, increase the mat border width so that the frame doesn't crowd the art

• Powerful pieces of art, either with bold colors or strong shapes, may require wider mat borders to balance them.

• A large focal point that fills the artwork may require wider borders to prevent it from looking crowded by the frame

• When a bright mat color is used as a predominant mat, consider reducing the mat border widths from what would be used with a neutral-colored mat

BOTTOM WEIGHTING

Bottom-weighted mats have wider borders below the art than on the other three sides. They help bring visual balance to a completed frame design when the art looks heavier in the lower half.

This shows the same print matted with different proportions. The upper design has uniform borders on all sides, and the width of the borders are modest. The lower example shows a more substantial mat size along with a vertical elongation, adding further to the design. ("Brown Butterflies" by Katrina Whitney. Courtesy of Artaissance, www.artthatfits.com)

This print is approximately 12"x16" so the scale of the fork is larger than life. Generous mat borders, combined with a wide moulding, help to build a strong proportional balance between the art and the frame design.
("La Fourchette" by Andre Kertesz. Copyright 2003 Ministere de la Cultre-France. Published by the McGaw Group, www.bmcgaw.com. Courtesy of Howard Greenberg Gallery, New York City.)

GUIDELINES FOR WHEN TO BOTTOM WEIGHT A MAT

• When colors in the lower half of the art are darker than the colors in the upper half, the art can appear bottom heavy. A bottom-weighted mat can help balance that effect

• When there are larger shapes in the bottom than in the top of the art

Art with perspective usually has larger shapes in the bottom foreground of the art.

• When there are heavier textures or bolder patterns at the bottom than at the top

• When there are low horizon lines in landscape paintings

• When framing valuable art that has a wider paper margin below the image

It is not advisable to alter the size of any valuable art. Doing so can immediately depreciate its value-now and in the future.

• To provide a vintage look for art since bottom weighting was commonly used in the past

• Increase the bottom border of the mat to add height so the framed art will work well in a tall, vertical space.

Bottom weighting mats is a traditional technique that has become popular again. Increasing the lower margin adds stature and also can help the art balance with a tall, narrow wall space.
("Salt Mine I" by Ed Kranick. Courtesy of Artaissance, www.artthatfits.com)

The amount of bottom weighting should be dictated by the art. Art that is extremely bottom-heavy calls for a mat with more bottom weighting than a piece that is only slightly bottom-heavy.

When valuable art has a wider paper margin below the image than on the top and sides, a bottom-weighted mat should be used. The amount of bottom weighting is dictated by both the paper margin below an image and by how much an image needs it.

The darkest colors and boldest shapes in this piece of art are in the lower half. Bottom weighting the mat (making the lower margin slightly wider) "lifts" the visual weight so that the completed design looks more balanced.
("Math 101" by T.L. Lange. Courtesy of Artaissance, www.artthatfits.com)

The texture of the basket and wavy shapes of the leaves in the lower portion of this art look heavier than the smooth flowers at the top. Bottom-weighting brings a better degree of balance.
("Calla Lily With Basket" by Greg Perkins. Private Collection of Greg Perkins)

Perspective makes things appear larger the foreground. In th case the fence keep getting larger as it reaches the lower portion of the art. Increasing the lower margin of the mat helps offset the perspective.
("Dune Fence." Copyrig 2004 Michael Kahn. Published by the McGaw Group, www.bmcgaw.com)

The mat borders on these two pieces are the same width. The bright persimmon is overpowering with this much showing, while the off-white looks great. Mat design can be improved by altering the borders depending on the intensity of the mat color.
("Sao Paulo Two," copyright 2004 Rex Ray. Published by the McGaw Group, www.bmcgaw.com)

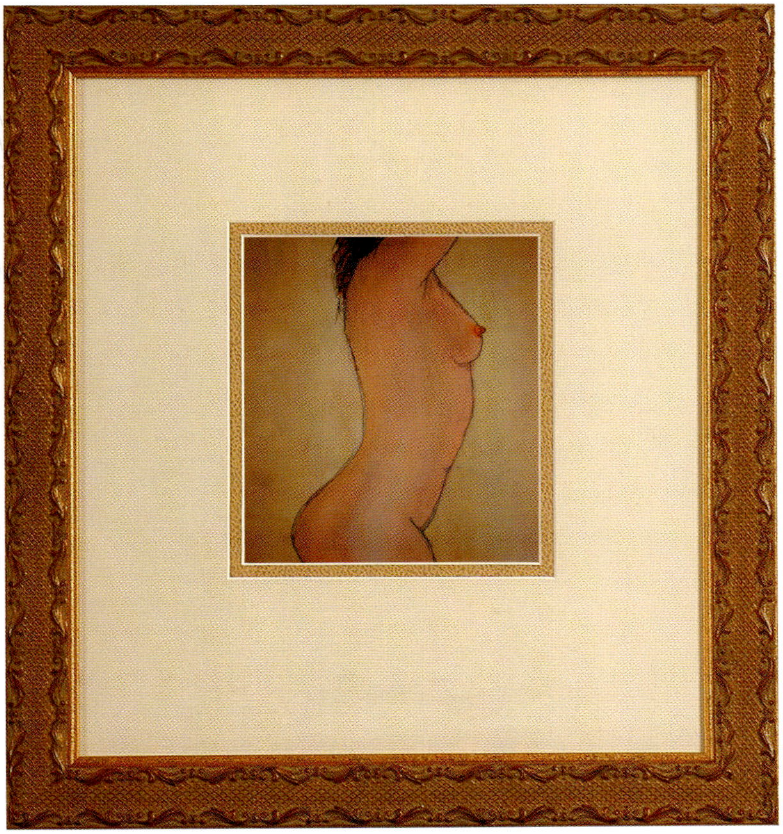

Using slightly bottom-weighted mats creates a classic look since bottom weighting was common in the past.
("Reach" and "Poser" by Greg Perkins)

Approximately 40% of the upper part of this print is light in color and pattern while the lower 60% is darker and bolder. A bottom-weighted mat helps create a better overall balance.
("In The Pine Woods" by Katrina Whitney. Courtesy of Artaissance, www.artthatfits.com)

Low horizon lines can create the appearance that the art is sinking into the mat. Adding to the lower mat border prevents this.
("Sunrise Over Canyon" by Pat Meras. Courtesy of Artaissance, www.artthatfits.com)

A dramatically bottom-weighted mat adds a unique sense of style to this frame design. This type of mat border was selected because the paper the art was drawn on is that size. On other pieces of art, you can use this type of mat simply because it looks good or to help create a sense of balance in a room with a high ceiling.
("Tomato" by Craig Alan. Courtesy of Deljou Art Group)

The exaggerated bottom weighting on this mat enlarges the small piece of art to ensure that it will be noticed.
("Fill in the Blanks I" by Greg Perkins. Courtesy of Artaissance, www.artthatfits.com)

Elongated mats or borders have long been associated with Asian art. In this particular case, elongation works well because of the vertical shapes and the diagonal positioning of the leaves angling up and down. Try elongating non-Asian pieces,too.
("Bamboo & Plantains I and II" by Chris A. Paschke, Designs Ink, www.designsinkart.com)

ELONGATING

An elongated mat is one that has two sides with wider borders than the other two. Asian scrolls have had elongated borders for centuries, but other types of art can take advantage of this technique if it enhances the design. Mats can be elongated either vertically or horizontally. I personally do not recommend horizontal elongation. Most people look at things from left to right, a habit picked up from reading. Wider borders on the sides tend to hold the viewer's attention. Alternately, when the eye reaches a narrower side border, it bounces back to the art, looking for more of the image.

GUIDELINES: WHEN TO ELONGATE A MAT

• ***It looks best on vertical pieces of art***

Elongation can also be used effectively on square or horizontal pieces when they have some of the following characteristics:

• ***Art with strong vertical lines***

Examples include picket fences or tree trunks.

• ***Art with vertical shapes or motion***

Examples include standing figures or balloons rising into the sky.

• ***Art that will hang in a narrow, vertical space***

OFF-CENTER MAT OPENINGS

Placing a picture off center in a mat isn't common, and there aren't many instances when it is even particularly flattering. But when it is appropriate, it can be highly effective.

GUIDELINES: WHEN TO PLACE A PICTURE IN AN OFF-CENTER MAT OPENING

• ***If an artist intentionally placed the image off center on the paper***

• ***If the strongest part of an image is in a corner***

• ***When framing a set of four prints that will be displayed two over two***

Place each image in a different corner of its mat so that the four images will be closer together when the grouping is hung on a wall.

This mat was vertically elongated to add stature to the already vertical print. Notice that the border is wider below the art than it is above due to the dark shadows in the lower portion of the art. This helps elevate the visually heavier bottom to create better balance in the completed frame design.
("Puerta Vieja I" by Doug Landreth. Courtesy of Artaissance, www.artthatfits.com)

By the time you add mat borders and frames around a set of artworks, they can push the individual pieces so far apart from each other that they lose continuity. In this case, offsetting each piece of art towards the inside corner pulls them closer together so they regain their effectiveness as a set.
("Abstraction", "Details", "Particular", and "Skeleton Back" by Jesse Canales. Published by the McGaw Group, Tribeca Works Collection, www.bmcgaw.com)

PRESERVATION

In addition to all the colors and patterns of mats that are available, mats are also made with different levels of quality. This book is about design, not preservation, but the basics of preservation are worth mentioning. The least expensive mats are acidic. They can damage art and reduce the longevity of frame designs. Have you ever noticed how quickly a newspaper yellows when sitting in sunlight for a few hours? This is due to the poor quality of the paper. Inexpensive mats eventually do the same thing. Even worse, they may also eventually ruin the art.

Better quality mats are made from cotton fibers or alpha-cellulose. These are the types of mats you should use when matting anything you care about and to protect your investment in custom framing. The important thing is to be aware of the options and use the best type of mat for each project.

GUIDELINES FOR USING PRESERVATION QUALITY MATTING

- ***Use it to protect anything having monetary value***
- ***Use it to protect anything having sentimental value***
- ***Use it to increase the longevity of the frame design***

This is a piece of original art. It has been matted and framed using conservation grade materials to preserve it. Even when framing inexpensive pieces, use conservation grade materials to insure the frame design will look its best for a much longer period of time.
("Untitled" by Pietro Adamo. Courtesy of Progressive Fine Art)

An off-center mat opening can change the overall balance of the frame design. In this case the darkest part of the art is way off in one corner. By offsetting the art in the mat, that dark area comes closer to the center of the completed design rather than pulling the viewer away.
("Crossing Over II" by Greg Perkins.
Courtesy of Artaissance, www.artthatfits.com)

This print is interesting but has no bold details to accentuate in the design. Simplicity is important to add interest to the framing without distracting from the delicate qualities of the art. A good look is achieved here with a combination of mouldings, 8-ply mats (to get an extra deep bevel), and spacers between the mats to add even more depth.
("Volute I" by David Dauncey. Courtesy of Artaissance, www.artthatfits.com)

ADDING DEPTH TO A MAT DESIGN

A single mat offers little in the way of depth. A bevel cut provides minimal depth, but there are alternatives that increase depth and add interest to a frame design.

Double or Triple Mats: Many people think of double and triple mats simply as a means of adding more color to a mat design. But the extra layers of matting also increase the depth of the presentation. If you use the same color for each layer, you will enhance the appearance of depth rather than focusing on the colors.

8-Ply Mats: 8-ply mats are twice the thickness of a typical mat. When an 8-ply mat is cut, it creates a much more dramatic bevel and accentuates depth.

Deep Bevels: If you want an even deeper bevel, you can create one with foamboard.

Once cut, the bevel may be painted, gilded, or wrapped with decorative paper. If a bevel is going to be painted or gilded, it should first be coated with gesso to prepare the surface. The gesso may need to be sanded to provide a nice finish. A gilded bevel can be glazed with paint or stain to create a patina, adding instant age. Craft-style acrylic paints are creamy and generally look better than artist's acrylics.

This picture was on a magazine cover, and it was quite tattered. Cutting around the art and removing the background eliminated the problem areas. Suspending it above a mat backing rather than mounting it flat creates quite a dimensional design.

An interesting way to add depth to a frame design is to recess art away from the mat in a deep frame moulding. Doing this also enhances the perspective within the art.

"Historical Personage" Original etching by Sarah Rishel. Private Collection of Greg Perkins)

Sandwiched Mats: Mats come with different color cores. When you combine more than one color, you create a sandwich effect. For example, you can attach a black core mat to a white core mat and cut the opening through both layers. This will create a stripe on the finished bevel. You can use three layers and create an "Oreo" effect. If you combine a 2-ply mat with a 4- or 8-ply mat, it can add both color and variation in depth.

Spacers: When you use double, triple, or quadruple mats, you can add spacers between the layers to increase the overall depth. This can enhance the perceived depth of a landscape or a still life. The depth of spacers added between the upper and lower layers of a triple mat can also be varied to create different depth effects.

Recess the Mat: You can increase the visual depth of a standard single mat by recessing the entire mat away from the glass.

Recess the Art: You can also keep a mat against the glass and recess the art farther back in the frame. When doing this, it is important that the role of the art isn't diminished by placing it too deep in shadows.

When small objects are recessed into a shadowbox, be careful that they aren't lost in the shadows. In this design, the mat border surrounding the button collection is placed near the back of the frame rather than up by the glass.

Spacers were used behind the mat and fillet, creating the space for the dimensional tiles. (Handpainted tiles from Studio Vertu)

CHAPTER SIX

GLASS & ACRYLIC

Many things that are framed should be protected with either glass or acrylic, which is called glazing. You may not think of glass as being a design element as much as a means of protecting what it covers, but it does contribute to design in more ways than one. When the various finishes of different types of glass are placed over art, they can make the art look sharp, soft, bright, or muted.

The quality of the glass will also impact a design's longevity. With regular glazing, the art and matting can discolor much sooner, reducing the integrity of the design. The additional UV filtering in conservation glass helps protect the value of the art and the overall integrity of the frame design.

In general, paper pieces like prints and watercolors should be covered. The exceptions may be if the art is laminated or has been coated with a clear acrylic finish. Typically, paintings on canvas and different types of needlework have not been covered with glazing, but there are exceptions. If the items are old and fragile; if they hang in a public area where they may be easily damaged from unwanted touching; or if there's a concern about smoke, grease, or other contaminants soiling them, glazing can be desirable.

When glass is used, one consideration is to keep it from coming in direct contact with the artwork. There should always be air space separating glazing from the item being framed. Matting will usually provide this space. But when a mat isn't part of a frame design, something else needs to create the necessary space. There are a number of different spacer products available to custom framers specifically for this use.

Different types of glass reflect varying amounts of light. Strips of three different type of glass are shown here. The Museum Glass in the center reduces reflection, making it easier to see the art. Museum Glass also reduces harmful UV rays to help protect the art.
("Port Clyde Main" by Ted Goerschner. Courtesy of Artaissance, www.artthatfits.com)

CONSERVATION

All glass protects art in the sense that nothing is able to directly touch it. This means it will not be damaged from being touched or by pollution or other airborne contaminants. However, light itself is one of the most damaging things to art. It can quickly ruin a nice piece of art and also degrade the frame design in much the same way that

sunlight damages human skin. To protect art as much as possible, be sure to use conservation glazing products. Also be aware that even conservation glass does not fully protect art. It is still smart to keep it out of direct, harsh light.

TYPES OF GLAZING FINISHES

Glass and acrylic products come in both conservation and non-conservation types as well as in a variety of finishes. Some of the finishes are:

- Regular Clear
- Non-Glare (also called reflection control)
- Anti-Reflective (doesn't glare and is clearer than reflection control)

These descriptive names vary from one brand of glass to another. Conservation products cost more than regular glass, but they protect the value of an item you are framing, so it is money well spent. Without it, you may end up spending more on new art and framing to replace the piece that was damaged.

GUIDELINES FOR CHOOSING GLASS

• Consider the lighting situation in the room where the framing will hang

Think about the time of day the room is used the most and what the interior and exterior lighting will be like on the specific wall where a frame will hang.

• Clear finishes keep colors and lines crisp and sharp

It is a good choice for bold, contemporary graphics or certificates with fine lettering.

• Non-Glare finishes soften lines and colors

They work well with soft focus art and photos and pieces with muted colors.

• Anti-Reflective finishes look almost like they aren't there

This is the best choice to avoid glare so you can enjoy your art more.

• It is perfectly acceptable to use different types of glass in one room

• When choosing glass, be aware of nearby windows and lamps. Anti-reflective glass or acrylic reduces glare without distorting like non-glare products.

Think about each piece of art and where it will hang.

This frame design includes three pieces of glass. In addition to the glass you can touch, there are two more layers behind it. The stamps are glued to those two pieces of glass as well as to the mat behind them. All of these stamps are cancelled so they have no special collector's value. This technique is purely decorative and should not be used when framing collectible stamps. In that case, the stamps should be placed in sleeves and may therefore look better placed under openings in a mat to hide the sleeves.

Photo by Clark H. Hill

CHAPTER SEVEN

MOUNTING

Nearly everything that gets framed can and probably should be mounted in some fashion. I use the term "mounting" very loosely and include hinging. Different types of mounting include both conservation and non-conservation materials and techniques. Items having either monetary or sentimental value should be mounted using conservation methods. When mounting something like a reproduction print that is both inexpensive and replaceable, non-conservation methods are fine.

WHAT DOES MOUNTING HAVE TO DO WITH DESIGN?

Since mounting is what goes on behind a picture or under a mat, you really don't see it in completed framing. However, it still can impact design. When a paper print is not permanently mounted, it can buckle from changes in temperature or moisture. Un-mounted papers can expand and contract with changes in temperature or humidity. Obviously, a buckling print can detract from the appearance of a frame design.

Also, the method chosen for mounting can make subtle changes to the overall appearance. For example, a shiny photo can be mounted in a way that prevents buckling. But that may change the surface texture of the photo slightly, which will be noticeable in its surface reflection. Or a piece of cross-stitch can be mounted with or without padding, resulting in either a padded or a flat look.

"Floating" is another mounting technique. To float a piece of art means to mount it in front of the mat. Sometimes floating also refers to raising the art off the surface of the mat. Floating allows an entire piece of art to show—all the way to its outer edges. This is especially effective when framing art with deckled edges or vintage pieces having chips or cracks that enhance their character and age. Floating is also important when framing something like a tile because it displays its dimensionality.

Art made with pastels tends to have a soft, velvety surface that can be easily marred. If the sharp edge of a mat sits on top of the art, it can leave an impression in the pastel that may show if it is ever reframed. To avoid this, the art can be floated in front of a mat. A second mat with a fillet in the opening serves as a border and also lifts the glass away from the art.

Photo by Clark H. Hill

When framing black & white photography, the most flattering designs are usually those limited to black, white, gray, or silver color schemes. That doesn't mean the designs have to be boring. What this Deco-style frame lacks in color, it makes up for with its interesting pattern, coordinating with the arch at the base of the Eiffel Tower.
("Under the Eiffel Tower I" copyright Linda Omelianchuk, wwwlindaomelianchuk.com)

CONSERVATION (A.K.A. PRESERVATION) MOUNTING

I will not get into the details of conservation except to warn you of its importance. When art is preservation mounted, it is typically hinged versus being permanently mounted. The idea is to make sure that all the processes are reversible, allowing an item to be returned to its original condition. This is important for resale of investment pieces. It may also be important for something like a child's painting that you may keep framed for several years then want to store it to save for years to come.

NON-CONSERVATION MOUNTING

Non-conservation mounting is done either with wet glue, spray adhesive, heat-activated tissues, or pressure-sensitive adhesives. Any of these will adhere the art to a mounting board. Typically, foam-board is used for this purpose, but sometimes other types of boards are used. This kind of mounting should be considered permanent and should only be used for inexpensive and/or replaceable items.

When framing original art, it shouldn't be cut or permanently mounted like a poster would be. In this situation, the art was hinged in front of a mat. This is called floating and allows the natural deckled edges of the paper to show. *("Buster II" by Scott Hile. Courtesy of Rosenbaum Fine Art)*

Tattered and torn edges may seem like flaws, but they can also add a lot of character to the frame design. This torn corner hints at the age of the map and lets the viewer know this isn't just another print but is an authentic antique map.

CHAPTER EIGHT

PUTTING IT ALL TOGETHER

The real trick to great frame design is being able to pull all the framing components together into a single, cohesive look that truly flatters the item being framed. You must use the perfect colors, style, and proportions. If any part of the design is out of sync, it can detract from the overall look. A beautiful design can soon become a lot less flattering if it isn't created with proper conservation materials. Whether you are framing valuable art or not, use quality materials to protect the frame design.

GUIDELINES FOR THE COMPLETE DESIGN

• Be aware of how all of the components relate to one another

Mats and mouldings that look great with art independently may not look good with each other.

• Adjacent components, such as the frame and mat, should be varied in width from each other

The eye flows more easily to the art when different proportions are used.

• Framed art should be used as an accent to add personality to a room

Avoid trying to match framing so perfectly to the decor that it almost disappears into its surroundings.

• Small details, such as a fillet inside a mat opening or an Enhancer in the lip of a frame, can take a frame design from so-so to fantastic

• Consider the space where the framed piece will hang

Plan the framing proportions to ensure that a framed piece looks its best in the allotted space. The same print can be framed differently depending on overall room size, length of a specific wall space where it will hang, ceiling height, and the scale of a room's furniture.

By choosing a frame, mat, and fillet from one color family, there is a subtlety to the frame design. The contrasting light neutral mat immediately surrounding the art pulls attention back from the ornamentation on the moulding so the art remains the center of interest. (Figurative by Ely. Copyright Deljou Art Group)

Photo by Clark R. Hill

#9

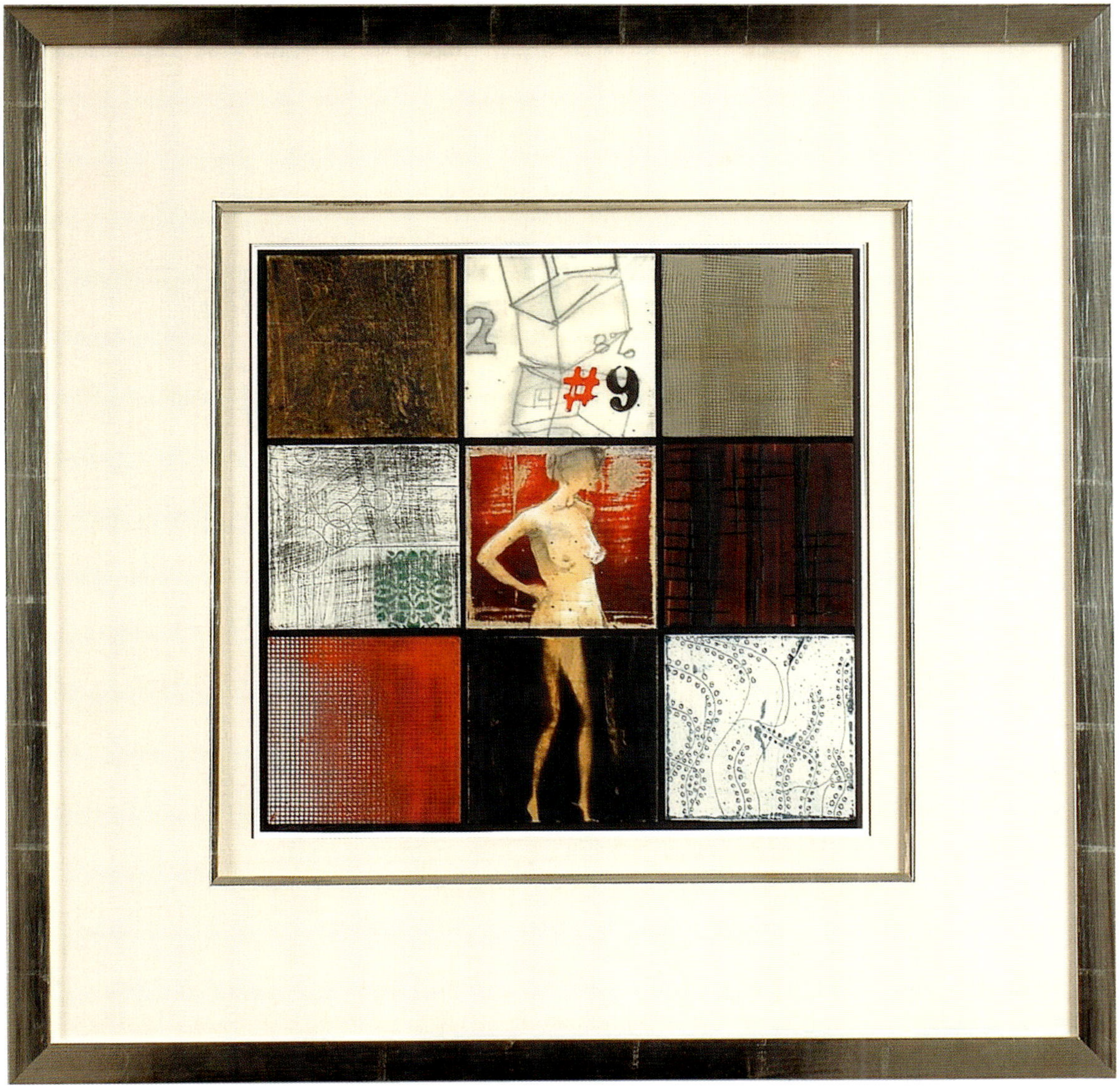

(Left) This design is all about squares. The artist combined nine individual square pieces of art into one presentation. Several of them have their own square patterns, as do the mats and moulding. All the squares in the framing work because they are subtle compared to the boldness of the art. This type of design would make a big statement in a room, so it may be perfect for a primary wall as a focal point in the space.

(Right) This frame design is much more subtle. The 12 karat white gold frame, linen mat, and fillet all work together to create an understated sophistication suitable for a fine art presentation.
("Writ in Form" by Craig Alan. Copyright Deljou Art Group)

Here a single frame is used. When you don't have a liner or mat, the frame may need to compensate for what is not there. This frame has interesting ornamentation, a contrasting lip and outer edge, as well as good scale.
("Sonnet" by Maeve Harris. Courtesy of Grand Image, Ltd)

In most but not all cases, it is desirable to create your frame designs so adjacent components are all varied in width from whatever is next to it. In this case the frame varies from the fillet, the fillet from the mat, etc. This makes it easier to focus on the art.
("The Heartbeat" by Marysia Burr. Courtesy of Grand Image, Ltd)

Brightly colored art often compels people to choose colorful mats and mouldings. Since this is a piece of original art, a subtler frame design helps maintain a look of value.
("Wide Awake #5" by Downe. Courtesy of Progressive Fine Art)

Botanicals are often the choice of interior designers for use in traditional spaces. This design has all of the elements of a classic frame design, including a water gilded frame, light neutral top mat with bottom-weighting, and a fillet. However, the striped under-mat adds a decorative element that gives it an updated look.
("La Bananier", "La Cocotier" and "La Palmier" from New York Botanical Gardens. Courtesy of Artaissance, www.artthatfits.com)

A cow may be a barnyard animal, but that doesn't mean it has to be surrounded by a barnwood frame. Using a more refined frame would allow this piece to be displayed in a more elegant setting.
("Dairy Farm I" by Dale Payson. Courtesy of Artaissance, www.artthatfits.com)

This photo contains both warm and cool colors. With age the paper has yellowed, making even the cool blue sky a warmer tone. In keeping with the current condition, a yellow-based green mat and a yellow-based tan mat were used.

SPECIFIC FRAMING PROJECTS

Art on Canvas

There are several options for framing art on canvas:

- **Frame Only**

When a single frame constitutes the entire frame design, that single element should be special enough to make up for the fact that there isn't a liner, a mat, or a fillet to add interest and scale.

- **Frame and Liner**

When framing a canvas, mats are not used but liners serve much the same purpose. Liners are often fabric-covered, but other frames can also be used as liners. When a liner is used, it takes some of the burden off the frame by contributing to the design.

- **Stacked Mouldings**

Combinations of mouldings can add color, style, or scale to framing. Flat or scoop-shaped profiles tend to look best as an inner frame. Nearly any moulding profile works as an outer frame.

- **Float Frames**

A float frame provides a means of presenting art so that none of the front surface is hidden. A float can be used as the frame, but it can also be surrounded by another moulding that relates to the personality of the art.

Art and Photos on Paper

Art on paper is typically matted. Mats provide a visual break between the art and the frame and also serve as spacers to keep the glass from touching the art. Many types of frames can be used. The best choice depends on the style of the art.

These three examples show how variations in the moulding selection alter the appearance of the completed design. All of the designs are comparable in color but the profile shapes provide a series of different looks. Even subtle changes help art look its best in the specific location where it will reside. ("Old Leather Chair" by Karen Lawrence. More of her work and contact info are at www.artshow.com)

Canvases aren't normally placed in shadowboxes under glass, but that can be a nice way to present them. It helps create unity between the set of three. Together the three small pieces have more impact. *("Asian Trilogy" by Greg Perkins. Private Collection of Greg Perkins)*

(Right) Sepia tone photographs include a narrow range of colors, limiting the mat and mouldiing colors that will enhance them. In this case, a mat that is close in color to the lightest color and a frame closer to the darkest helps create a sense of contrast similar to that in the photo.

(Above) The same photograph, this time printed in a slightly different color, is framed very differently. An ornate frame provides a more classic look. This style works equally well with the picture but is more appealing to people who like a traditional style.
("The Violinist" by Linda Omelianchuk, copyright Linda Omelianchuk, www.omelianchuk.com)

(Top) Sometimes a float frame is used as the sole frame. However, some float frame profiles may be too basic to bring out the personality of the art or balance its visual weight.

(Middle) Even a frame as simple as this gold profile adds enough that it gives a more finished look. It is a good choice for a minimalist presentation.

(Below) This bold, exciting painting has the strength to stand up to a much bolder frame. A black and gold frame balances with the rich colors used by the artist and the scale and depth of the frame's ornamentation work well with the artist's heavily textured brushstrokes.

(Original painting by Deanna Sirlin. Copyright Deanna Sirlin)

The rustic fence and the hair on the dog were the inspiration for the distressed wood frame and soft suede mat in this frame design. If the dog was in a more elegant setting, the frame could go that direction, too.

The pewter color of this frame highlights the colors of the photograph. Because the frame, mat, and fillet are all in mid-tone grays, the dramatic light and dark colors of the photo remain strong in the completed design.
("Winter in France" copyright Linda Omelianchuk, www.omelianchuk.com)

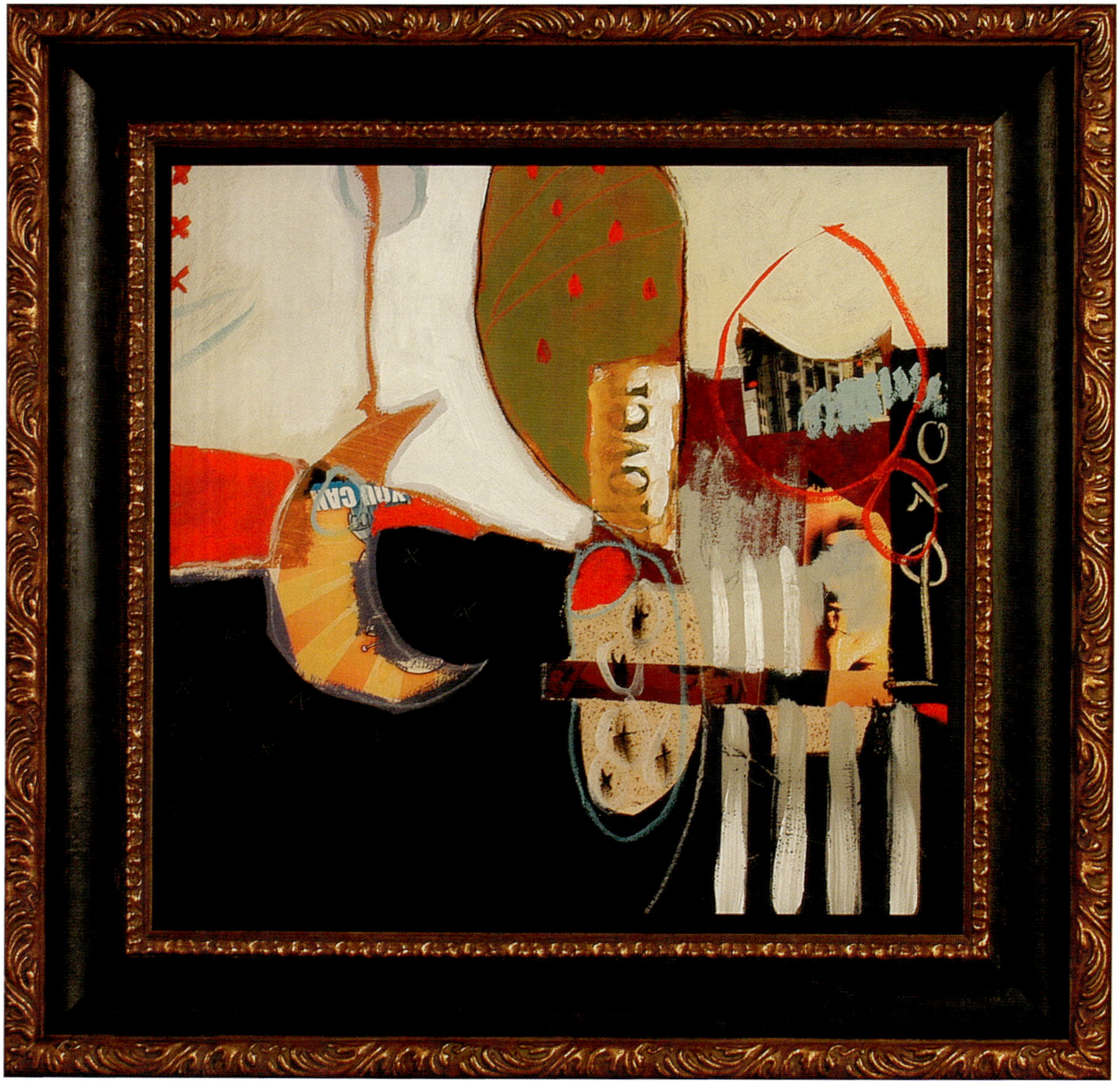

If you can't afford original art, frame a poster to look more like a painting. A substantial moulding helps this poster look like a much more expensive piece. In this case, the white margin surrounding the image was cut away and the art floated over a black mat.

("Canal Street" by Tucker. Copyright 2004 DAG Inc., Derek Tucker. Tribeca Works Collection, published by the McGaw Group, www.bmcgaw.com)

POSTERS

Posters are usually the least expensive type of art, which is the reason why they're often framed inexpensively. The results are generally uninspiring, and they become more temporary than other framed art. You can upgrade the perception of a poster with better framing, which will also make it more appealing and likely to be kept a longer time.

A poster can be matted to enlarge its visual impact or to add color. If a poster has its own border surrounding the image, consider covering that border with a mat to avoid redundant borders.

Posters can also look more important when a stacked moulding combination is used in place of a single, narrow frame. You can also add a splash of color with a liner or perhaps inject a pattern using a frame with ornamentation.

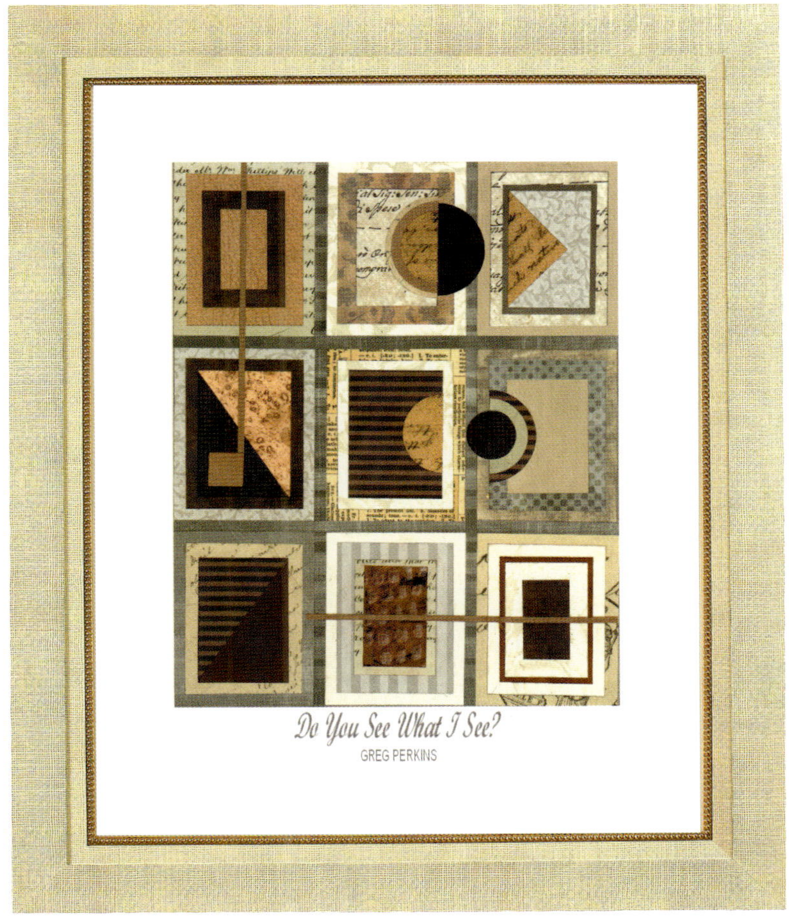

You can customize the framing for your favorite posters just as you do for other art. Take a close look and you will see the the outer frames shown here have a linen look that relates to the squares and lines in the art while the beaded fillet ties into the circular shapes.
("Do You See What I See?" by Greg Perkins. Print of this image available from Artaissance, www.artthatfits.com)

Just because a poster is a poster doesn't mean you have to choose an inexpensive "poster" frame to surround it. Here, a stacked moulding combination elevates the look of the poster to that of a higher quality piece of art.
("Africa by Air" by James. Copyright 2004 Brian James. Published by the McGaw Group, www.bmcgaw.com)

Small posters can become interesting pieces of art when they are framed creatively. This design includes a triple mat and a two-tone frame with the color scheme working hand in hand to offset the poster.
("Samba D'Orpheus New Orleans" original artwork by Joanna Kreisel. Prints copyright Wild Apple Graphics)

When all of the colors in the art are bold, neutral mats remain the best color option. Also, the light airy mat border allows for this elongated design without being overbearing to the art.

("At Last I" and "At Last III" by Greg Perkins. Courtesy of Artaissance, www.artthatfits.com)

DRAWINGS AND ETCHINGS

Drawings and etchings are generally quite delicate, with fine details that could easily be overpowered by a strong frame design. Both types of art have little visual weight, so the framing should typically be lightweight-looking, too. If you have trouble with the concept of visual weight, think of how a glass of water looks lighter in weight than a glass of chocolate milk—even though they are very similar in physical weight. Frame designs for art with little visual weight should typically use light colors, fairly narrow mouldings, and small-scale frame ornamentation—if ornamentation is used at all. If wider or bolder frames are desired, keeping them light in color and minimal in contrast to the art will help the design.

If you fear that the overall frame design will be too small, keep in mind that you can use generous mat borders to increase the size. If the mat color closely matches the paper that the art is on, the lines in the art will tend to project and stand out on their own.

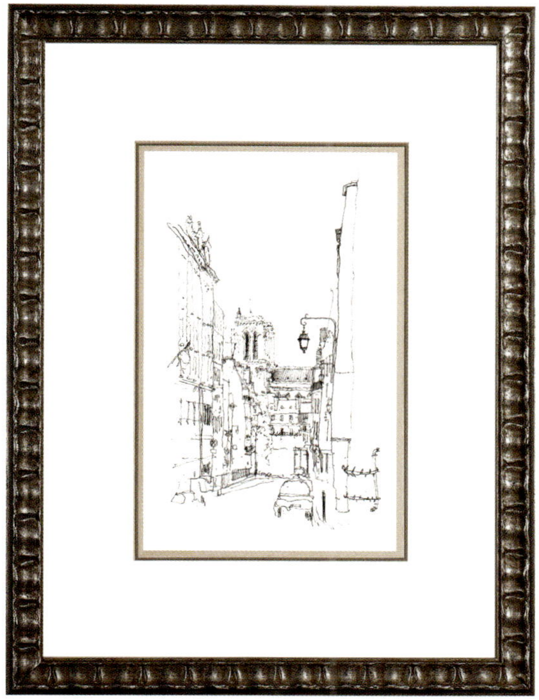

This line drawing could easily be overpowered with too much color or pattern. At the same time, not doing enough cheapens the overall look. This frame works well because its pattern isn't overly detailed and the monochromatic triple mat pulls attention into the art. *("Rue Valette, Paris" by Mat Barber Kennedy. Courtesy of Artaissance, www.artthatfits.com)*

The delicate features of this frame and fillet coordinate with the fine details in the sketch. There is no bright white in this artwork, so a light gray mat was used to help call attention to the off-white figure. *("Contemplating" by Greg Perkins)*

All types of collections can look wonderful when framed. These arrowheads gain importance when clustered together and placed in a generously proportioned moulding. The silver Enhancer inside the frame has a pyramid pattern that coordinates well with the points of the arrowheads.
(Arrowheads. Private Collection of Greg Perkins)

Sometimes you can take what you want to frame and use it as part of your frame design. In this design a carpenter's ruler was used to decorate the mat.

Even the most unusual things can be framed. Unexpected objects become interesting conversation pieces and they can add a lot of personality to your home. These pottery shards, found on a trip, become a great souvenir from that event.
(Pottery Shards. Private Collection of Greg Perkins)

OBJECTS

Object framing is a broad category and one that is more difficult to describe. Differences in the objects as well as the wide variety of framing materials and techniques allow nearly endless design options. Basically, a frame needs to be deep enough to contain the object(s) being framed along with glass, mats, and backing. Shadowbox or object box mouldings come in various depths up to about 4". If additional depth is required, one moulding can be placed over another to extend it to the necessary depth.

In general, objects are mounted to a background mat. They tend to look best on fabric mats, such as linen, silk, or suede. Contrasting fabrics tend to set off an object so it shows up well. For example, soft suede contrasts with coins or plates while silk contrasts with a velvet purse.

The same framing techniques used for paper prints can also be used when framing most dimensional objects. This includes using mat borders and fillets. After all, the objects worthy of framing usually have sentimental if not monetary value and should be framed to reflect that.

Possessions that are passed down from one generation to the next may be perfect candidates for framing. If you haven't inherited heirlooms worthy of framing, antiques stores, estate sales, and flea markets are filled with interesting things you can frame to personalize your home.
(Courtesy of Scott Howell)

The way objects are arranged within a frame can be as important as the actual frame and mat choices. When you take the time to think through an interesting layout, it will be much more enjoyable to look at the completed design in the years to come.

Kalagas are a hand-embroidered art form found in Thailand and Burma. They are heavy because there are thousands of beads, stones, and sequins adorning them. A mat could be used as a border around a Kalaga but a liner provides more stability for the art. Also, the added depth of the liner helps visually support the textile.

NEEDLEART AND TEXTILES

After the countless hours a person spends stitching a piece of needlework, it should be properly framed for it to look its best for years to come. The starting point is proper laundering so that oil from the skin as well as any other soiling won't detract from the piece when it is framed. If you aren't an expert in this area, find someone who will do the job properly. It may also be necessary to block needle art to re-square it. Again, rely on an expert.

Various types of needlework are framed differently. A cross-stitch is often matted and covered with glass. It is much less common to mat a needlepoint. When a border is desired for a needlepoint, a liner is frequently used.

It takes hours and hours to stitch even a very small piece of needleart like this one. When someone invests their heart and soul into a piece of needleart, it is worth a lot. Here a delicate French frame was used and a contrasting fillet added for its light color.

(Above) Although the materials for this project didn't cost much, the labor invested into it makes it is worthy of a nice frame. A small, simple design like this would be a perfect choice if it were going to be displayed on an easel within a small bookcase or in other confined spaces.

(Right) This is the same cross-stitch but with a frame design that includes mat borders. The larger size helps it fill a larger space.
(Sampler by Nancy Perkins. Private Collection of Greg Perkins)

If a quilt is beyond repair, frame the salvageable pieces as a keepsake. If a quilt is passed down in the family, sometimes people deconstruct it so each member of the family receives a piece. Rather than making a pillow, framing provides a great way to display it.

Family heirlooms are wonderful things to frame. Once they are on display, you can enjoy them every day and the framing can help protect them for future generations.

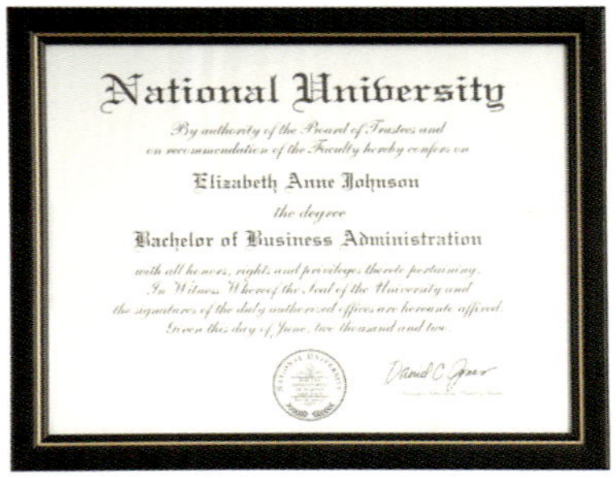

(Top left) For most people a diploma is the most expensive piece of paper they will ever own. Don't settle for a cheap "diploma" frame like this one when you can express your personal style with a multitude of much nicer choices.

(Lower left) A nicer moulding with a furniture finish and a generously proportioned double mat help the same diploma look more polished and professional.

(Top right) In this design a veneered wood frame is lined with an Enhancer for added detail.The mat is accented with a black fillet, tying in with the color of the Enhancer.

(Lower right) Since a diploma usually costs more than all of the art a typical person will ever purchase in their lifetime, why not treat it like a work of art and frame it as such? This design uses a water-gilded frame and a double black linen mat with a fillet between them. This classic design is one that could proudly hang anywhere.

DIPLOMAS AND CALLIGRAPHY

A college education is one of the most costly purchases people make. That means a diploma is more expensive than any of the art you are ever likely to buy. The price of that sheet of paper makes it worthy of a special frame, not just an inexpensive "diploma" frame.

Diplomas, like paper art, should be matted, glazed, and framed. Some people choose their school colors for the mats, but neutral matting generally adds longevity to the frame design and offers more flexibility for displaying it in different places over its lifetime.

Don't feel obligated to use a predictable black frame. Burl wood, gold, or a number of other choices can look great. Avoid trendy frame designs that will become outdated since you may want to display your diploma throughout your career.

Other types of word documents, including calligraphy, can be framed using all of the same products and techniques as diplomas, Because calligraphic works can also take on a much more artistic flair, more creative touches may also be used.

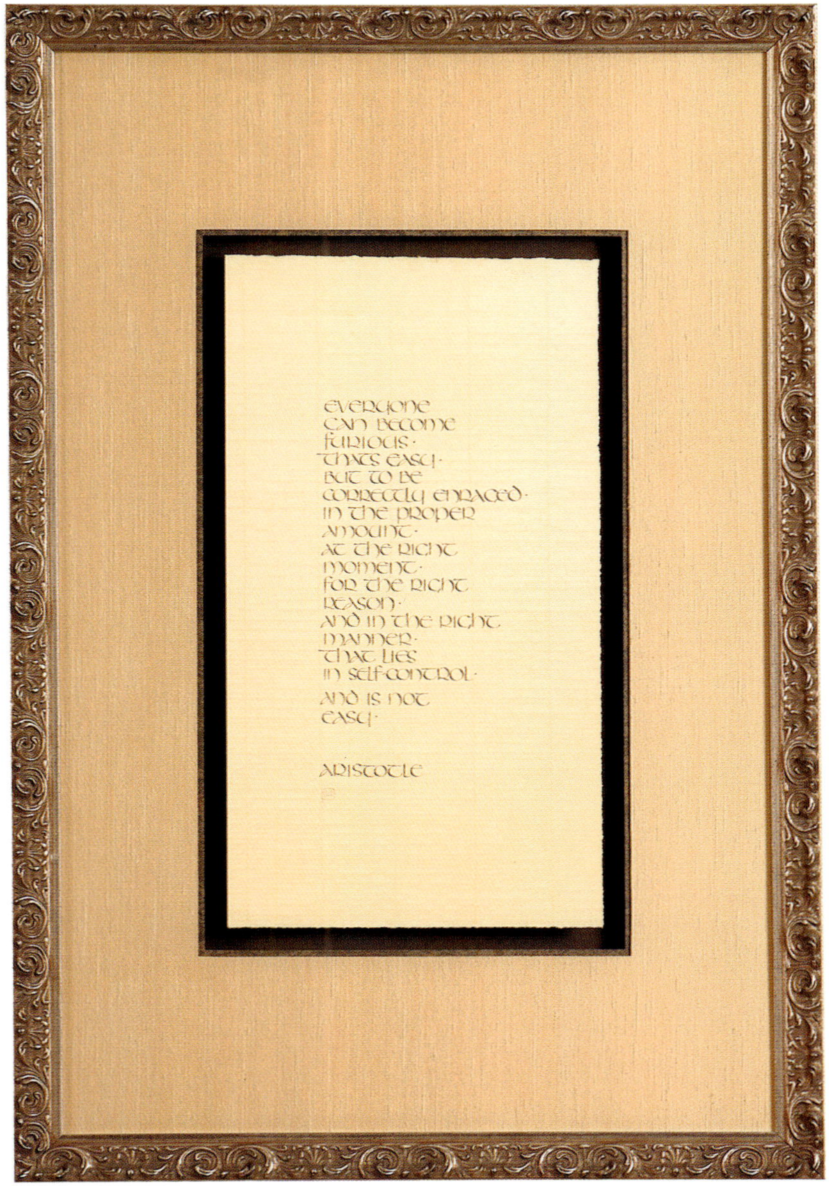

This calligraphy is so finely detailed that framing could easily overpower it. In this design, fine ornamentation on the frame relates to the lettering and a dark mat behind the calligraphy pulls attention back to the art.
(Calligraphy by Chris A. Paschke, Designs Ink, www.designsinkart.com)

You can bring a bit of the landscape indoors by positioning mirrors to capture part of your outdoor view.

Both natural light and light from lamps are reflected into this mirror to help brighten the entire room.

Mirrors expand space. This reflection makes it seem as if you're looking into another room.

All photos by Clark R. Hill

MIRRORS

Because a mirror has no inherent style of its own, it provides a rare opportunity to use any style, color, and width of moulding. It makes a lot of sense to choose something that will relate to room decor. It can either blend in or make a grand statement by standing out. Mirrors aren't just for checking your hair and makeup. They also can pull light into a dark corner, visually enlarge a space, or help balance a room by reflecting color or interest from the opposite wall.

Framed mirrors fit well in any room in a home. Beautifully framed beveled mirrors are commonplace in bathrooms these days, where the metal-frame medicine cabinet was once the star. And those tall, thin mirrors in cheap plastic frames that were mounted to the back of closet doors are now becoming an integral part of room decor, with nicer framing. The main consideration when framing mirrors is supporting their weight.

Mirrors are backed with silver to make them reflective. Therefore silver frames tend to blend with mirrors. Gold frames typically add a sense of formality. Deep rich wood tones look traditional, and white frames are great for a beach house or little girl's bedroom. The point is that mirrors are versatile and allow you to customize them to work with any decor.

Photo by Clark R. Hill

Leaning mirrors rather than hanging them provides a nice alternative.

Mirrors can be used as trays under collectibles or flowers on a dining table.

LIVE
MAINE
LOBSTER
WATERMELON

CHAPTER NINE

LOOKING FOR SOMETHING UNIQUE?

One of the wonderful things about custom framing is that it offers the ability to customize. If you simply want to fill an empty corner, you can use basic products and techniques. If you are looking to create a real focal point for a room, custom framing can help you achieve the right look. The next few pages show some out-of-the-ordinary frame designs that can open your mind to options you may not otherwise think of.

This "ladder" of framed artwork can be a great solution if you are lacking wall space. It will lean against the side of an armoire or even a refrigerator. This design consists of three fully framed pictures and two 7' long pieces of frame moulding to form the ladder's legs.
("Kraft I, II, & III" by Scott Hile. Courtesy Rosenbaum Fine Art)

The extremely tall mats provide a great backdrop for the baskets of bamboo. Framing like this makes a big statement in a room.
("Enlightenment" and "Spiritual Freedom" by Chris A. Paschke, Designs Ink, www.designsinkart.com)

Photo by Clark R. Hill

If you are looking for a one-of-a-kind way to fill a large wall, look beyond a single large piece of art or even traditional groupings. This grouping consists of 16 identical frames. Eight of them contain original calligraphy of Asian characters. Two hold mirrors. Four frames do not have glass in them, leaving the deep shadowbox frames accessible as shelves to set vases in. This same type of grid grouping could be altered by using family photos with something belonging to each person in the adjacent frame.
(Art by Chris A. Paschke, Designs Ink, www.designsinkart.com)

(Above left) Save the corks from wine bottles to create a unique serving tray. Glue the corks to a dark backing board and place in a frame that is deep enough to accommodate the corks, glass, and the backing. You can add handles, such as drawer pulls for kitchen cabinets if desired.

(Above right) Many of the objects people frame are important family heirlooms, significant collectibles, or other equally meaningful possessions. No matter what the reason for framing it, it is worth considering all of the same special techniques and details used for good pieces of art. In this example, mats, a fillet, and a special base to anchor the object on make this a very unique shadowbox.
(Carved Wood Figure. Private Collection of Greg Perkins)

(Left) Three small pieces of art make a big impact when they are placed in dramatically bottom-weighted mats. The large expanse of the mat below the art creates a nice backdrop for the flowers. You could substitute mirrors for the art and sculptures for the flowers. These use the same design concept shown on page 119.
("At last I," "At Last II," and "At Last III," by Greg Perkins, courtesy of Artaissance, www.artthatfits.com)

Although these four prints have different colors, they are a matched set of art. To unify them on the wall, use the same mats and mouldings on all of them. Ornate frames add drama to the contemporary art.
("Shifting Sands I, II, III, & IV" by Sean Jacobs. Courtesy of New Era Portfolio, www.newerahd.com)

CHAPTER TEN

DISPLAYING FRAMED ART

Once your custom framing is complete, the next important step is to display it. There aren't hard and fast rules regarding display, but there are some tips to help your pieces look their best.

GUIDELINES FOR DISPLAYING FRAMED ART

• Don't place large or visually heavy pieces over small, delicate pieces of furniture

Large-scale framed art will look bulky and cumbersome when paired with furniture that isn't comparable in scale.

• Don't place too small a piece over a large piece of furniture

It will look insignificant and out of scale and will be dominated by the furniture.

• The best place to see custom framing is at eye level

Keep in mind that "eye level" varies for each person's point of view, so compromise may be necessary. Also, eye level is different in a hallway, where you stand to look at framed art, versus a dining room, where you are seated most of the time.

• When displaying frames directly over furniture, keep them relatively close to create a unified presentation

If the separation between art and furniture is too large, the two components won't appear to relate to one another.

• When choosing where and how high to hang something, be sure to consider the surrounding accessories

Don't allow framed art to be hidden by a plant or be too close to a candlestick on a mantel, for example.

• Carefully arrange groupings for visual balance

Formal rooms may look best with symmetrically balanced groupings whereas asymmetrical displays look appropriate in more casual settings.

• Art hung in unusual spots is becoming more accepted and common

For example, a grouping can be hung from floor to ceiling beside a tall bookcase, or a small piece may be displayed under a table in an entry hall.

• Place art on a large easel to fill an empty corner

In large rooms, furniture is often pulled away from the walls to create more intimate seating areas. This may leave an empty corner that can be filled by art.

• Lean and overlap frames to create vignettes on a tabletop or mantel

Homes with many windows, open floor plans, or built-in bookcases may not have sufficient wall space for an art collection to be displayed. Lean several pieces together on a tabletop, even slightly overlapping them.

• When framing matched sets, match the mats and mouldings throughout the grouping.

• When grouping diverse art images, frame them differently to reflect the style, color, and personality of each piece

• When framing a collection of similar items (such as family photos from different generations) you can bring cohesiveness to the grouping by using different styles and widths of moulding, all in the same color family. Or you can use two or three colors from the same style of frames.

A FINAL NOTE

Custom framing has the power to both enhance and protect art, objects, and a variety of other items. Compared to other items you can purchase, custom framing is truly one of the best values available. When you take the time to make smart choices that result in quality frame designs, they will endure for many years. Your initial investment is just pennies a day.

Remember, an investment in custom framing can become a special family heirloom that you can pass along to future generations. Whether your budget allows you to purchase original art or you are limited to reproduction prints, both can be framed to look like a million bucks. Good quality framing can, indeed, transform a home or office into something spectacular.

An accent mat can be the perfect place to introduce more color into a design. A colorful top mat is a major commitment to a color, but a small amount of color peeking out from a neutral mat brings more versatility to the design. The soft green used for this accent helps draw attention back from the frame to the artwork.
("Buchoz Leaves" by Vision Studios.Courtesy of World Art Group)

Combining subtle patterns and textures in this stacked moulding combination adds a high level of sophistication to this design.
("Casa de Campo" by Luis Fernando Hebrero. Courtesy of Artaissance, www.artthatfits.com)

This painting sits in a black float frame, which in turn is surrounded by a wider gold frame. The float presents the art so none of the surface is hidden under the frame. The classic styling of the gold frame adds a dramatic touch to the otherwise contemporary art. *(Abstract fashioned after an original painting by James Leonard by Greg Perkins. Private Collection of Greg Perkins)*

Framing a wine bottle can commemorate a special event like a wedding, an anniversary, or even a trip to the wine country.

Refine the appearance of a distressed wood frame by pairing it with a more elegant Enhancer. Combinations like this result in the casually elegant look that is so popular.
("Stick Horse #4" by Jaime Ellsworth. Courtesy of Artaissance, www.artthatfits.com)

This style of frame may seem like an unexpected choice for this art but, when you study the art and see all the wavy lines in it, both the frame and Enhancer inside of it make sense.
(Untitled by TL Lange. Courtesy of Grand Image, Ltd.)

INDEX

TYPES OF FRAME DESIGNS

Frames in Museums 7
Frames in Galleries 9
Framing for Corporate Use 9
Custom Frame Shops 10
Framing for the Home 10

FRAME DESIGN GUIDELINES

What is a Guideline? 13

GUIDELINES FOR MOULDING

Color 16
Style 23
Width 29
Depth 29
Liners 32
Float Frames 35
Fillets 37
Enhancers 39
Stacking 41

GUIDELINES FOR MATS

Colors 47
Pattern/Texture 59
Borders, Uniform all Sides 65
- Bottom-Weighted 67
- Elongated 73
- Off-Center Openings 73

Preservation Matting 75
Adding Depth 77

GUIDELINES FOR GLASS

Choosing Types of Glass 83

GUIDELINES FOR COMPLETE DESIGN

Putting It All Together 89

GUIDELINES FOR DISPLAYING

Displaying Framed Art 121

FRAME MOULDING

What is a Frame Moulding? 15
Moulding Color 15
Moulding Style 21
Moulding Scale 27
Liners 30
Float Frames 35
Fillets 37
Enhancers 39
Stacking Mouldings 41

MATTING

What Is a Mat? 43
When Should a Mat Be used? 43
Mat Colors 47
Mat Pattern/Texture 59
Borders, General 63
- Uniform on All Sides 65
- Bottom Weighting 65
- Elongating 73
- Off-Center Openings 73
- Preservation 75

Adding Depth to the Mat Design 77

GLASS

General Information .81
Conservation .81
Types of Glazing .83

MOUNTING

General .85
Conservation .87
Non-Conservation .87

PUTTING IT ALL TOGETHER

The Complete Design .89

FRAMING SPECIFIC THINGS

Art on Canvas .95
Art on Paper .95
Posters .101
Drawings and Etchings .105
Objects .107
Needleart and Textiles .109
Diplomas and Calligraphy .113
Mirrors .115

UNIQUE FRAMING TREATMENTS

Looking for Something Unique? .117

DISPLAYING FRAMED ART

General .121

All photos by Clark R. Hill